MISSION COMMAND

in the

Israel

Defense

Forces

Edited by Brig. General Gideon Avidor

UNG

UNIVERSITY *of*
NORTH GEORGIA™
UNIVERSITY PRESS

Blue Ridge | Cumming | Dahlonega | Gainesville | Oconee

Published by:
University of North Georgia Press
Dahlonega, Georgia

Printing Support by:
Lightning Source Inc.
La Vergne, Tennessee

Cover and book design by Corey Parson.

Cover image by Israel Defense Forces, licensed CC BY 2.0.

ISBN: 978-1-940771-76-2

Printed in the United States of America

For more information, please visit: http://ung.edu/university-press
Or e-mail: ungpress@ung.edu

UNG
UNIVERSITY of
NORTH GEORGIA™
UNIVERSITY PRESS

Blue Ridge | Cumming | Dahlonega | Gainesville | Oconee

Table of Contents

Part 4: Mission Command Put to Test

Part 5: Mission Command Over the Horizon

Appendix A

Introduction

Brig. General Gideon Avidor

Command and Mission Command

The dictionary definition of "command" presents variations on the theme of "authority"; however, the idea of "command" is much more complex. According to the dictionary, command can only exist when other people carry out the instructions issued by a commander. Indeed, command involves not only authority but also action.

Military command generally focuses on a mission with clearly defined objectives that must be achieved through coordinated activities. Because the commander is responsible for achieving these goals, command also involves responsibility.

Command (both authority and responsibility) has been demonstrated throughout history according to a variety of techniques. In the last decade, the "mission command" method has become the most popular mantra or buzzword.

Preparing an army for war is a process that takes many years. It involves building the force and developing and training officers and soldiers for battle. Each army does so according to its own heritage and in accordance with its own national and local culture.

Like any other army, the Israel Defense Forces (I.D.F.) has dealt with mission command over the years in its own unique way. The present volume will demonstrate how, for the I.D.F., mission command is not merely a technique, but a culture.

Shaping the I.D.F.'s Battle Doctrine

When an army prepares its officers, a significant gap occurs between slogans and methods and reality. In some areas, this gap is objective—a result of lack of knowledge, uncertainty, or the inability to know the enemy sufficiently. In others, it is subjective and due to the army personnel themselves.

In most regular armies, the accepted procedure is that the battle doctrine is written by experienced officers in the army's headquarters and distributed by the general staff to units for their application. This doctrine is updated periodically or as the result of an exceptional occurrence, such as during the impact of the 1973 Yom Kippur War.

Armies are constructed hierarchically. Large organizations, like armies, who engage in such complex activities as warfare cannot function properly without a hierarchical structure and discipline. This is true of the I.D.F., but we are also a growing army lacking a historical tradition.

This has its advantages and disadvantages. Despite being constructed to run along normative lines, the I.D.F. is unique, as will be elaborated on in this collection. For now, we will state that "unofficially, the I.D.F. promote[s] decentralized command, also called 'mission command'" (Finkel, 2014).

This approach is based on multiple assumptions. The battlefield is fraught with uncertainty, and often the best solution is to afford maximum independence to junior officers. This thought is then furthered by the premise that these officers have the best knowledge of the mission and therefore will do whatever is necessary to complete it successfully (Finkel, 2011). In 2013, the I.D.F. officially proclaimed mission command to be its preferred command style.

Over the years, the I.D.F. has fought in six conventional wars (in 1948, 1956, 1967, 1973, 1982, and 2006) with the participation of large formations such as brigades and divisions. Between those wars, the I.D.F. has continually confronted terrorist organizations and faced escalating tensions along Israel's borders. These activities were generally on a low tactical level of limited operations nature, including operations in Judea, Samaria, and the Gaza Strip (during 2002, 2008, 2010, and 2014). The I.D.F. refers to these operations as "campaigns between wars" and considers them ongoing threats that must be dealt with in a manner that is a total departure from conventional warfare.

There is a fundamental difference between commanding an army unit during a war and the required preparation, training, and other activities "between the wars," when there is no meaningful enemy present. During "between the wars" training, the effort invested in supervision, management, and accompanying tasks is greater than that devoted to big formation operations.

The commander's responsibilities in periods of preparation are more restricted than in wartime. During preparations, they are surrounded by countless supervisors, managers, staff officers, and advisors and are required to explain their activities almost as much as the objectives they want to achieve. In battle, the commander is measured almost exclusively by their achievements and is rarely called upon to offer explanations about the ways they operate—except after failed operations. Thus, we often encounter a gap between a commander's behavior at routine times versus on the battlefield. Every army faces the problem of narrowing this gap and successfully training commanders in peacetime to function in wartime.

In calm periods, the hierarchical, bureaucratic military system functions in full force. In wartime, the situation is different. The transition from routine to war constitutes a major difficulty, one that requires time to overcome as efficiently as possible. Commander and forces training intend to ease the process; without that, it will be difficult to apply mission command on the battlefield. There might be isolated cases of exceptional commanders who successfully manage to make such application, but the command and control system will not be capable of sustaining it over time if not prepared in advance. Thus, despite long periods of routine and limited operations, the I.D.F. attempts to inculcate mission command principles, even if it does not always apply them.

The Israel Defense Forces came into being as a people's army "in motion." It had its beginnings in the midst of a war (1948) based on concepts of underground armies, partisans, and Special Forces. The functional abilities of its officers in battle were (and are) beyond anything else due to the training and performance throughout their careers. Traditionally, I.D.F.'s battle doctrine has been determined on the field. A maneuver or idea that succeeded in the field was subsequently presented to the General Staff and established as compulsory doctrine. Alterations to doctrine also originate in the field, as new requirements or methods engender temporary doctrine that later made permanent.

Since its foundation, the I.D.F. has developed and advanced commanders based on functional-operational abilities rather than academic or intellectual ones, which are important in themselves, but not obligatory. Officers who lack a high educational level will acquire it during their army service with the I.D.F.'s help.

Attaining commander status in the I.D.F. is based on the principle of rising up through the ranks. This means selecting soldiers for command positions based on their proven abilities at lower level—one step at a time—and on their evident desire to pursue a military career. This system is a departure from the accepted practice in most of the world's armies, which fill officer ranks according to selection processes that take place before recruitment.

Attaining commander status in the I.D.F. depends on one undergoing proper training based on active service. From the lowest levels, the system for selecting commanders, again, heavily relies on proven ability: a section commander will be chosen from among the best soldiers; an officer will be chosen for advanced training from among the most successful petty officers; an officer will be promoted based on their performance "in the field." Senior commanders must serve as junior officers in relevant positions beforehand; a battalion commander must first have commanded a company. Every promotion and every additional training course will depend on proof of ability, unrelated to family background, social class, education, or ethnic roots. Everybody begins at the bottom.

From the beginning, the I.D.F. adopted mission command as its preferred approach. It taught mission command in all command-training courses and emphasized it in every drill, even if it's not always applicable in practice. Commanders are expected to act according to this principle, but in low intensive operations or those operations subject to public opinion—and most military activities between wars fit those categories—the higher echelons might limit it. In the commanders' reports presented below, such tensions frequently occur.

In wartime, the transition to mission command is natural and continuous. Very few I.D.F. commanders will wait for instructions from above when faced with a battle situation. A striking example of this may be found in the memoirs of Yoram Yair, the commander of the 35th Battalion during the First Lebanon War in 1982 (Yair, 1990).

Mission Command in the I.D.F.

Mission command is a command and control strategy within the broader context of leadership. The present volume deals with this method as it applied in the I.D.F. Like any topic that falls within the province of behavioral sciences and deals with interpersonal relations, mission command cannot be defined mathematically or isolated from the culture of the society in which it is practiced. Because of the vast differences between how it is applied in different countries, the I.D.F. model is fundamentally different from the American, German, British, or Australian one.

These differences stem from education toward excellence at home and at school, where the schools, colleges, and universities expect the students to surpass the teacher. The inculcation of this approach in the I.D.F. begins in junior officer training courses and is widely discussed (mainly in periods between wars) until it is well known and accepted as the norm. Every commander strives to realize this goal, if only they are allowed to do so; if they are not hindered on the way, it is the natural path they will choose. An accepted starting-off point among junior I.D.F. commanders is that, in their kitbags, there is a general's baton and, given the chance, they will prove the appropriateness of this gear. Until then, they will follow orders and do their best to successfully complete missions as ordered by their immediate superiors.

This educational principle states that the commanding officer will determine their intention and dictate the mission, together with the limitations and contingencies imposed on them by the higher ranks. If not instructed otherwise, the junior commander will act according to mission command. They will not wait for instructions from above and, even when given such instructions, will frequently suggest improvements to the original plan or suggest a plan of their own. Such principles are evident in the texts that appear below; whether written by commanders who have experienced active service or by military theoreticians, they all deal with the nature of mission command, how it is applied, and in which situations.

Mission Command and Leadership

Mission command and leadership are inseparable. Whereas leadership deals with the individual's ability to inspire people to carry out their commands, mission command deals with applying leadership when activating a unit or a formation. Leadership might be applied in many ways and styles according to the leader's personality. The question of whether this leadership is inborn or acquired by learning is irrelevant to our discussion. Since it takes place within an operational organization, military leadership includes authority and responsibility for performing missions.

Since expectations from commanders are absolute in wartime, they tend to use their authority to direct subordinates' activities, keeping control over them "to avoid mishaps." Subordinates expect to receive clear, unequivocal orders according to which they plan their next move, with very narrow margins in which to do so. Mission command widens these margins considerably, while still relying on authoritative and professional leadership; indeed, it cannot be effective without leadership.

The mission command approach places the responsibility for carrying out a mission on the shoulders of the lower level commander, as they are the most conversant with conditions and local opportunities. If the commander possesses the qualities necessary to carry out their plans, and the motivation to act in the best manner for achieving their goals, they should be supplied with the ways and means to do so in their own way.

The senior commander is required to clearly define their intentions, ensure that their subordinate understands them correctly, enable their junior officers to carry out their plans, and support them in doing so. This does not lessen the senior commander's responsibility, as the junior commander's mission is on their shoulders, and the commander must ensure that it is fully carried out. The commander does this through supervision of battle developments and supportive intervention if necessary. For this approach to work, there must be a close, trusting relationship among commanders at all levels.

The senior commander must remember that the emphasis is on carrying out the mission and that any plan is a good one as long as it overcomes its weak points and reaches its objectives. Any problem has more than one solution and commanders should choose the best one, no matter who suggests it.

Mission command is also directed at easing mental pressure on the commander. This leadership style entails "loosening the reins," while giving junior commanders the sense that they are trusted and can function according to their own ideas. This can elevate their motivation and dedication to the mission to new heights.

Mission command demands a high level of professionalism, leadership, and mutual dependence among command teams. Consequently, intensive commander training must take place before mounting operational missions. Many prior conditions are necessary to produce good commanders, including familiarity with battle doctrine, command and control theory, and military jargon. Commanders must be well-acquainted with and trust the subordinates and soldiers under their command. They must demonstrate patience with juniors and a willingness to absorb their mistakes. The Israel Defense Forces have been grappling with these issues since its establishment. It was founded in an unconventional manner, and its spirit persists to the present day.

What Does the Future Hold?

Since I.D.F. commanders rise through the ranks in a prolonged track based on operational performances, they have already acquired considerable insights and experience when they arrive at the higher command levels. Mission command accompanies them the entire way, and their expectations from subordinates and their behavior in any framework is influenced by this approach. In addition, when circumstance causes them to act independently, mission command is always present in their deliberations.

As long as a tradition continues of shaping battle doctrine during operations and afterwards by the designated staff centers, the spirit of mission command will continue to prevail. Like any army, the I.D.F. demands routine activity, including planning that takes place according to fixed rules, but, when deemed necessary, the establishment will support junior officers acting according to mission command.

In the I.D.F., mission command is a well-rooted cultural tradition. It is difficult to apply in peacetime, but it is still in everyone's consciousness and supported by the chain of command. When it is relevant to circumstances, it is the natural, accepted solution.

The Present Volume

This book presents mission command in the I.D.F. from a variety of standpoints during various periods, including commanders' deliberations about how to apply it in specific situations. We present the topic through the eyes of I.D.F. commanders at multiple levels of seniority and from different generations. We can see differences in adaptation to conditions at different periods, and can participate in their dilemmas, but the central track has always been to strive toward mission command as the preferred leadership style.

In the background stands the army framework encouraging mission command on all levels. The connecting thread running through this collection is that the insights are from combat leaders in the field, as they are more interested in finding practical ways of applying this approach rather than analyzing it theoretically. The theory came second, as a necessary response to explain what had occurred in the field.

The book is divided into four sections:

- Section 1 describes the theory of mission command in the I.D.F. in the eyes of scholars. It contains three academic studies and one study on theory and practice from a former Chief of the Navy.

- Section 2 provides historical views on the struggle between concepts that were part of the I.D.F. build-up efforts during its early days. These ideas include Western militaries' regulations versus Special Operations improvisation and more.

- Section 3 moves from the theoretical to the practical, with commanders giving their insights based on their field experience. These commanders include first-star offices at the GHQ level with a wider perspective than that of second field commanders, who derived their understanding from current field experience.

- Section 4 explores mission command in battles, including real life senarious from commanders, staff, and warriors fighting in real life battles.

- Section 5 looks to the future and discusses mission command in the Information Age.

Part 1

Theory

Part I

Theory

1 Command Systems and Control

Combat Leadership and Ground Forces Mission Command

Dr. Uzi Ben-Shalom

Introduction

On the battlefield of ground forces, use is made of a growing number of advanced technologies, and this trend is likely to increase (Kott, 2008). In the framework of this process, extensive use is made of online communications systems as an aid to the process of command and control (C2 systems).[1] The introduction of C2 systems to military organizations is only part of a wider process, which in the past was designated "a revolution in military affairs." This refers to the technological modernization of the battlefield, particularly the development of advanced weaponry and advanced sensors providing information from the battlefield and displaying it as computerized and integrated. It has been widely predicted that the increasing utilization of such information systems and sensors will disperse "battle fog," thereby minimizing uncertainty, and the commander will possess reliable information that will enable them to make better decisions.

Despite these predictions, the battlefield remains as uncertain as ever (McMaster, 2009). Furthermore, based on writing about innovations and technological advancement at various periods, it appears that the existence of advanced technology does not change the basic nature of fighting as a human activity involving violence, killing, randomness, and embarrassment (Marshal, 1947; Keegan, 1975; Sinnreich, 2008). The war

---

1   Henceforth, C2 systems; these systems have been developed from C2 (command/control) systems to combined C4i networks (command, control, communications, computers & intelligence).

of the future, even if extremely advanced technically and technologically, will still not be a "war of buttons" or a "clean war." As long as human beings are involved in the act of war, leadership will remain a vital basis for armies, as this factor enables the commander to control their men and reach their objectives (McCoy, 2006). Generally, the discussion of mission command focuses on conveying orders, understanding the battlefield, and decision-making methods. In the present article, I will suggest an additional—equally important—direction and will claim that the ability to act in the spirit of mission command, that is, according to the commander's understanding of the mission and their officers' intentions in changing circumstances, is closely connected with combat leadership. This stems from the fact that activity in the spirit of mission command relates to both the emotional and the personal relationship between commanders and their subordinates—and among themselves.

I will also claim that combat leadership cannot be explained by means of qualities alone, but that it is also influenced by the manner in which human beings actually communicate with one another. The increased use of C2 systems bears on this subject, since changes in the field of communications are likely to influence the frequency of encounters, means of communication, its content, and the relative position of the participants. In addition, such communication systems demand expertise in operating them, reaching conclusions, being capable of deriving meaning from data, and making decisions—not only regarding the forces' activity or lack of it, but also while taking into account the emotional state of those under command.

A survey of the literature reveals that, compared to aerial forces, ground forces place greater weight on the human element of war than on advanced technology (Johnson, 2002). That being the case, a basic question arises regarding the ability of ground force commanders to utilize innovative C2 systems. This is because integrating such systems might clash with traditional conceptions of leadership or command on the battlefield. Furthermore, various aspects (both technical and practical) must be formally learned before an attempt can be made to implement such systems.

In the present article, I will attempt to point out a few such matters based on a case study of the I.D.F.'s ground forces. It is based on the current literature, conversations with commanders, and reflections on the significance of integrating C2 systems into the military leadership course taught by me at the I.D.F.'s College of Tactical Command of Ground Forces. I will not refer to a specific system but will deal with the topic generally, while referring to the various ground corps and units up to the formation level.

Combat Leadership and Mission Command

Leadership is one of the commander's major roles. When assuming command, they must serve as a leader of their men and influence them. This emotional influence is clearly reflected in their readiness to go "above and beyond" what is necessary (Bass, 2008)—in their preparedness to exert extreme effort and take risks. These voluntary expressions stem from trust (in goals, commanders, friends, and weaponry), in conviction of being well-prepared, and self-confidence. These aspects are heavily influenced by the commander's leadership concept in the eyes of those commanded by them.

Leadership can be defined in a variety of ways, but in the military context, it must be defined in relation to the concept of command, since military leadership and military management are two means of actualizing military command (Malone, 1997). Military command is a process in which the commander gathers information, processes it, formulates a decision, and issues a command which must be carried out. The command might take a number of forms, including verbal instructions to an officer or team or a written and detailed operational order to a formation. The military commander assumes various roles, including processing information, making decisions, and transmitting leadership. Whereas information is processed by the commander and their staff, leadership remains the sole province of the commander.

Military leadership means the ability to motivate subordinates to perform missions. It is a tool for creating discipline through the subordinate's identification with the commander and the commander's ability to impose his will by means of the constraints in his possession. However, direct leadership and personal ability are not enough; leadership also relates to the commander's professional abilities. A commander who is not a professional will find it difficult to gain the trust of their soldiers, trust that is so vital on the battlefield. In times of war, the commander's professional expertise is put to the test more than any other factor.

Leadership tasks are strongly linked to technological systems as these systems inform the leader of the combat situation and enable them to derive meaning from such information. Through them, the leader is also able to transmit and receive orders, maps, and diagrams. Therefore, the military leader must understand these systems in depth; their control of these processes are observed by others and taken into account when they ask themselves if they can trust the leader during times of stress. In addition, the employment of advanced technological and communications systems influences matters of command and control, while also enhancing responsibility, freedom of action, decision-making techniques, deciphering data, and deriving meaning from computerized data. All of the above are extremely important in the context of the leader's level of education and his ability to lead professionally.

Leadership that widely employs advanced technology—especially interactive computer systems, internet, and control at a distance—will formulate various unique characteristics. This topic is close in spirit but not identical with leadership "from close up" and leadership "at a distance," a gap widely discussed by commanders (Malone, 1997). The use of technology is likely to create a difference in not only the structure of conceptual knowledge but also the qualities of the leader himself. One striking influence relates to patterns of communication between the leader and those led by him. Another influence is the possibility of "dividing leadership," as subordinates are able to acquire knowledge that in the past was the exclusive province of the formal leader. Furthermore, various subordinates can communicate among themselves, exchange opinions, and be mutually influenced.

Challenges Presented to Commanders by Technology

I will now deal with some central matters that present challenges to military leaders in the ground forces in the context of utilizing C2 systems.

The Challenge of Training

Research in the field of leadership and C2 systems in the world are still in their early stages, and it appears that the possibility of basing educational programs on case studies is still limited. The existing training is mainly technical and does not encompass the dimensions of these systems technically (for example, broadcasting methods) and philosophically (for example, the sources of the RMA revolution and its links with uncertainty). However, the ability to exploit the best of sophisticated combat systems depends on the depth of training afforded to the officer. The officer must be given the opportunity to acquire understanding of the system through first-hand experience, including demonstrating how it operates, its disadvantages, and how its advantages can fully be utilized. For this purpose, the officer must be capable of understanding technology and processing information while also enlisting leadership in support of challenges of command and control. Military leaders seeking to implement C2 systems in support of their command must strive toward a professional view of the subject from the technical and theoretical standpoint while remaining cognizant of its impact on communication patterns between the leader and his men. This being the case, the concept of combat leadership today must undergo certain changes; it cannot exist without relevant technical training—as leadership ability depends on this—as opposed to only on establishing a model of the qualities that the leader should or should not possess.

It is extremely important to investigate the ways of relating to C2 systems in the context of current fields of activity, as we possess only a paucity of information from historical sources. In the I.D.F., most experience in this field has been acquired through ongoing security activities, but this is also true of Western armies, most of whose activities are carried out opposite lower-level semi-military organizations. Perhaps in the future, it will be possible to draw conclusions from the Russian Army's fighting in the Ukraine, but it may be claimed that the Western armies have not fought against an equal enemy for more than a generation. The Israeli example in this sphere is important and relevant. One instance of this is the fact that ground force commanders' experience has been acquired in low-grade fields of action. Commanders base part of their personal professional truth on experience acquired on such a battlefield, in the course of which they were required to derive significance from data (Ben-Shalom, Klar, and Benbenisty, 2012). In my conversations with commanders on this topic, they frequently mention the great importance of their own personal experience, most of which was acquired through ongoing security activity.

However, in this context, habits and personal experience can also be an obstacle. Since operating C2 systems in ongoing security activity is generally effective regarding continuity and does not suffer from interference or interruption, there is no need to deal with fighting processes, such as formulating a picture of the situation or analyzing complicated battle incidents. In ongoing security, the abilities of the enemy are generally limited, and there is no attempt to interfere with C2 systems. At the same time, C2 systems can cope successfully with the position of one's own forces. The limitations of the ongoing security context are magnified by the I.D.F.'s inherent leadership code, which emphasizes activity and experience. As difficult as such incidents might be, they cannot present a complete picture of fighting against a maneuvering enemy. However, since the "true" learning of the systems is currently taking place during ongoing security, information regarding activating C2 systems in maneuvering battles is liable to be only partial.

An additional important training area relates to basic skills, including orientation, managing manual equipment at headquarters, map reading, and navigation. Displaying symbols on a digital map is not sufficient to understanding what is actually occurring; however, employing systems that include these abilities mechanically is liable to decrease basic skills and the acquaintance with work methods that had been previously effective. In my conversations with commanders, they emphasized that C2 systems cannot totally replace basic skills. Despite the existence of advanced systems, soldiers must still learn to fight using those reliable old means. As one battalion commander expressed it:

> A good commander must possess a good understanding of computers and communications, which are a condition for fighting over the net. He must understand the technical side, but must also relate to the social significance of the system. By his nature, a commander wishes to be involved, so it is necessary to deal with this during the training process. Part of the knowledge that the commander should receive is when not to employ the system, when he should rely on "sticking his head outside" and receiving information from his surroundings—for example regarding fighting spirit and discipline in battle. Another aspect is when to employ the system and for which purposes. For example, it is worthwhile discussing with commanders when and when not to intervene in their subordinates' activities.

Deriving Meaning from Data

The command and control system influences the ability to derive meaning ("sense-making") from data and information. This concept offers a broader spectrum than does "awareness of the situation" when one is acting in an emergency. It refers to a series of active cognitive processes that come into play when an individual or a team become aware of a gap between their present knowledge of what should occur and what is actually happening (Klein, Moon, and Hoffman, 2006). The purpose of these processes is to reconcile this gap and adjust their understanding of the situation to reality. For this purpose, a variety of data must be collected and considered according to their cognitive significance. This process takes place in a variety of contexts, including military operations (Ben-Shalom, Klar, and Benbenisty, 2012). The ability to make sense of data and information through employing C2 systems, especially in emergencies, depends on habit and prior experience (Weick, 1995).

The mysteries of technology and the theoretical knowledge of military concepts inherent in weapon systems are not a part of what is required to turn soldiers into commanders in the I.D.F. And a commander who makes use of such systems is liable to derive only partial benefits from what they have to offer. For instance, in order to obtain maximum benefit, it is not sufficient to be capable of employing computer applications; it is also necessary to be aware of threats to the communication and delivery system supporting the system. In other words, it is difficult to plan a battle or conduct it without being aware of the limitations of technology.

Furthermore, C2 systems were constructed and developed assuming full knowledge of the wartime fighting situation. This assumption can rarely be totally realized, despite the presence of more data and numerous and sophisticated collection systems (McMaster, 2003). This being the case, commanders' ability to lead and command by such means depends on their ability to formulate general assumptions and be aware of their limitations. Through constant study and hands-on experience, they must formulate a professional command position regarding the use of such systems.

An additional important aspect of deciphering information displayed by the system is the ability to understand what is relevant and what is not. Digital systems can rapidly convey huge amounts of information, but not all of it is significant. This is another sphere of sense making, which mainly consists of separating the electronic "wheat from the chaff." Commanders must be able to comprehend what is relevant out of a wide range of data from various sources. It is very easy to become a slave to that material. Thus, when dealing with commanders' professional ability, it is worthwhile providing them with large amounts information, only some of which is relevant, thereby enabling them in developing the ability to critically relate to what is displayed before them.

Communication Between Commanders and Subordinates

Young military leaders report for army service from a rapidly changing social world, and a large number of their subordinates share with them habits of using advanced telephone and communications systems. The younger generation is no stranger to communications systems and digital media, but the older generation of commanders might view this component of leadership as an obstacle.

The chief role of leadership, especially in pressured situations, is to radiate a sense of security. This central role of the leader in the estimation of his subordinates has profound and varied psychological roots. Since they limit direct contact between the leader and their followers, C2 systems are liable to create a gap between personal, direct leadership and leadership from afar, which is unclear and uncommitted by its very nature. An example of this situation can be found in the following quotation from a letter relating to the role of electronic mail that was written by Maj. General (Ret.) Yitzhak Brick, the I.D.F. ombudsman, who was formerly a prominent field commander and veteran of the Yom Kippur War and decorated with the Badge of Honor:

> On a visit to one of the brigades, the commander told me about an incident in which he instructed one of his staff officers to prepare an order for a certain professional activity involving danger to life. After a while, it became clear to the brigade commander that the officer had not gathered the involved parties, had not briefed them and had not given them a clearly organized order. All the officer had done to prepare for the activity was to convey instructions to those involved via electronic mail. When the brigade commander became aware of this, he stopped the process and chastised the staff officer. The latter did not understand why he was being reprimanded, as he had been behaving thus for a very long period. This is one of a number of examples of inappropriate use of electronic mail, which in certain cases is liable to place soldiers in unnecessary danger. Electronic mail is a valuable and convenient administrative tool, but it is important to consider how it is used. It might possibly use as an efficient, but not exclusive, means of refreshing commands, but it is doubtful if commands and instructions should be issued by means of it. A commander who makes use of E-mail for conveying commands does not look his subordinates in the eye or conduct a dialogue with them, thus contributing to a cutting off interpersonal relations between the commander and those he commands. Furthermore, in this manner both the organized military practice of receiving and processing commands and the trust placed in commanders is damaged. (Brick, 2010)

The above extract illustrates a case of confusion between leadership and administration. The ability to make contact electronically is not a replacement for leadership in its most basic form, which involves exerting a direct influence on soldiers and commanders. This is especially true regarding missions that soldiers or their commanders are not enthusiastic about performing. In such missions, especially in unclear conditions, it is essential to ensure that the commander's intentions are carried out. In fact, an important fundamental of battle leadership at all levels involves considerable personal investment; the existence of C2 systems is liable to distract the commander from that principle due to the creation of "electronic distance" between the commander and his men, thus weakening his leadership skills.

The research literature in the field deals with measures of "heat" and "cold" in communications technologies. The use of C2 systems defines the transition to a "colder" medium than those used in the past. In other words, it is a tool that is liable to blur the leader's emotional messages. Communication via electronic mail is "colder" than communication involving the commander's voice, whereas voice communication is "colder" than face-to-face interaction between the commander and their soldiers. In this sense, C2 systems are liable to create a psychological distance between the two. This might have even more significance in situations of pressure or threat and danger, since in such situations it is vital for those led to experience the commander's spirit directly in order to sense their self-confidence and perceive their intentions.

Indeed, when commanders relate to leadership in the context of command and control, they are very aware that electronic mediation is liable to cause a distance between themselves and their subordinates. Every one of them expresses the position that command/control systems are insufficient and that it is necessary to maintain constant contact with secondary commanders, especially when problems arise. Their main point is that a ground forces combat commander's professionalism enables him to comprehend the psychological sensitivities and needs of the men fighting for him and with him. This is important to remember when so much information is displayed before commanders, since the human dimension cannot properly be conveyed electronically and can only be perceived through direct communication.

Another important element of leadership is the ability to be personally present at any crisis or problem location. C2 systems are capable of widening the area in which the ground battle takes place. This is one of their tremendous advantages; instead of crowding together, it is possible to spread out. Instead of being dependent on physical contact with the units, it is possible to describe or imagine them, thus improving the units' chances of survival and activating them more efficiently. However, in order to reap these benefits, leadership is necessary based on wide professional knowledge—both mutual and independent—which can only be derived from comprehensive training.

Centralism

Military leadership is one of the components of military command and has strong ties with the command philosophy of ground forces. "Mission command" is the epithet for the Israeli ground forces' official approach to decision-making during battle. According to this approach, the battlefield usually is characterized by communication problems and an accumulation of mishaps. This situation demands a distribution of decision-making to the proper echelons. Mission command provides an answer for this, as it is based on a commander's freedom to act at their level and with a readiness to take risks (Storr, 2003). A survey of the literature and commanders' comments indicates that employing C2 systems is definitely liable to present challenges to the "mission command" approach—i.e., the ability to distribute responsibility—since using such a system involves centralism, the avoidance of risks, a tendency to circumvent rank, and a slowing down of operational performance (Sowers, 2008).

The ultimate use of command and control must be based on acquiring experience, understanding the systems' influence on leadership and formulating an appropriate conception of it. In fact, it appears that commanders experienced in employing command and control demonstrate better distribution of labor when dealing with information. For example, the staff head might be responsible for processing all information arriving at headquarters, thus freeing the commander to go out and exert their leadership, especially if the situation demands it.

Limiting the Commander's Decision-Making Ability

The existence of online information on the battlefield and transferring it to various ranks are liable to limit the commander's ability to make decisions and exert leadership. For example, a formation commander who formerly served as a company commander is liable to assume that he has better understanding of the situation on the battlefield than does the current company commander. An inherent danger to this might result in too much involvement and circumventing levels of command, thus damaging leadership in the lower echelons. From the point of view of the lower ranks, C2 systems are liable to fortify the presence of "big brother" and increase an atmosphere of lack of responsibility, thus damaging the willingness to assume initiative and take risks (Larsen, 2009; Singer, 2009). The influence of such phenomena is liable to create a diversion from accepted military doctrine and create a different type of command dependent on authorization from above (Ben-Shalom, 2011). This is illustrated in the following comments by a company commander on a training course:

> In commanding nowadays, tension created between involvement and interference. If the command concept is mission command, then freedom should give to carry out missions. Command and control systems make it easy for the commander to negate this entirely. Today, each time you stop by the roadside to order a pizza ... somebody sees you. Today, I am certain that many times soldiers "trip over" wires, disconnect them because they are hungry, and want a snack. In such

a situation it is difficult—very difficult—to be involved, but easy to interfere. Take mobile phones, for example. Once you would order someone to travel from one place to another. Today he is constantly reachable. At one time, if he did not want to do something, he simply did not do it.

Final Comments and Reflections

Ground forces in all the world's armies have been quick to absorb C2 systems; these advanced systems are directed at making the commander more efficient and increasing the fighting power of formations maneuvering on the ground. Many consider this process from the point of view of technology and weapon systems. They acquire equipment, vehicles, and digital applications supporting each of the endless functions included in these systems. However, the success of such systems might seriously be affected by various elements, especially in the field of leadership. Combat leadership has a direct connection with mission command, as it enables its fulfillment and promotes the ability to fight in the spirit of its orders and missions. Such leadership enables the spirit of mission command when there is no possibility of making contact with the lower echelons and in the context of unavoidable difficulties.

In this chapter, I attempted to emphasize the influence of C2 systems on military leadership in light of a process of absorbing such systems in the I.D.F.'s ground forces. I am indeed convinced that the Israeli case is unique, but it appears that training in C2 systems in the context of leadership is still very limited in other armies as well. In this context, the Israeli case is also important for other armies. Similar to comments voiced in the past in the U.S. Army, the I.D.F.'s commanders are divided in opinion concerning the transition to C2 systems (Eisenstadt and Bacevich, 1998; Ben-Shalom, and Shamir, 2011). Some of them support this process and view it favorably; others are cautious and raise various doubts regarding its effectiveness. More experience will gradually be acquired in operating these systems, while their ability to function efficiently in extensive ground battles will be put to the test. However, their possible effects on leadership must also be taken into account.

The use of C2 systems has a profound effect on work methods and such important combat skills as navigation or conveying reports. However, thorough training in using these systems is currently being acquired in a limited fighting environment that does not place strain on them. Finally, commanders are obviously aware of the gap between electronic and interpersonal interaction; they are also cognizant of the vast difference between information displayed on the screen and verbal communication as a means of spurring forces into action. At the same time, there are undoubtedly cases where such a gap is created, thus presenting a challenge to leadership.

The Persona of the Digital Commander

Leadership might be linked with the commander's formal authority, their ability to impose sanctions or exert indirect influence. The commander's awareness of the officer's traditional role is an important component of their ability to project leadership to their subordinates. According to the British military historian, John Keegan, the commander does this by means of a "command mask" that they assume and by which their soldiers view them (Keegan, 1987). Keegan claimed that this mask is formulated not only based on the political and cultural structure of the nation and its army but also according to the nature of missions and available weapons systems—in the present case, C2 systems. The mask of command is a basic factor enabling mission command, as it represents the basic human elements of men at war: hopes and aspirations; requirements and dangers; love and justifications for sacrifice—and all of these combined with the cornerstone of mission command, namely, taking calculated risks in unforeseen circumstances according to a grasp of the situation. Combat leadership is influenced by C2 systems and, in turn, influences the manner in which commanders employ it. Until enough experience is accumulated in the use of these systems, there is definitely the possibility that they will present a challenge to commanders' leadership abilities and might create a distortion in the link between leadership, administration, and command.

While C2 systems improve the ability to process information, control the situation, and make decisions, they might also adversely affect some basic leadership principles, in the chain of command in general and the senior command in particular. It is worthwhile paying attention to the distinction between "the battle maps"—the picture viewed by the senior command—and artillery and the "contact battle"—the field as it is perceived by the lower fighting ranks coming in actual contact with the enemy (Akavia, 2006). C2 systems have a profound influence on this matter, as they generally increase a tendency toward centralism and intervention in the functioning of junior commanders. At times, they are accompanied by a slowing down of decision-making. Combat leadership that is capable of dealing with these influences must include both professional knowledge relevant to this field and an awareness of possible influences on colleagues and subordinates.

From all of the above, it appears that technological training for commanders should be broadened in the context of their experiences as activators and leaders. This should take place while deepening knowledge of the foundations of war and the proper place of advanced technologies on the battlefield. Although this cannot be considered a revolution—as C2 systems are merely a technical means of supporting processes with which commanders are already familiar—it definitely constitutes an all-encompassing change that is only at its inception.

Bibliography

Abd al-Muneim" Maktaba al-Suruq al-Dawliyah, Cairo.
Akavia, G. "Fundamental Barriers for the Use of C4I Systems in the

IDF." (Maarachot, 407, 2006), 17–29. (Hebrew).

Bass, B. M. The Bass Handbook of Leadership (4th ed.). (New York: Free Press. 2008).

Ben-Shalom, U. Trends in Training for the Military Profession in Israel—The Case Study of the Tactical Command College. (Political and Military Sociology: An Annual Review, 42, 2014), 51–73.

Ben-Shalom, U., Klar, Y. & Benbenisty, I. Characteristics of Sense-Making during Combat. (In M. D. Matthews & J. H. Lawrence (Eds.), the Oxford Handbook of Military Psychology. NY: The Oxford University Press. (2012), 218–231.

Ben-Shalom, U. & Shamir, E. *Mission Command between Theory and Practice: The Case of the I.D.F. (Defense and Security Analysis 27, 2011), pp. 101–117.*

Brick, Y. "I.D.F. ombudsman letter concerning the use of e-mail as a tool for sending orders to subordinates." (2010).

Eisenstadt, M. & Bacevich, A. J. Knives, Tanks, and Missiles: Israel's Security Revolution. (Washington Institute for Near East Policy, 1998).

Harig, P. T. "The Digital General: Reflections on Leadership in the Post-Information Age." (*Parameters*, 1996, Autumn), pp. 133–140.

Keegan, J. The Face of Battle. (London: Penguin. 1975).

Keegen, J. The Mask of Command. New York: (Penguin Books. 1987).

Klein, G., Moon, B. & Hoffman, R. F. "Making Sense of Sense Making I: Alternative Perspectives." (IEEE Intelligent Systems, 21(4), 2006, pp. 70–73.

Kober, A. "The Intellectual and Modern Focus in Israeli Military Thinking as Reflected Armed Forces and Society 1948–2000." (Ma'arachot, 30, 2003), pp. 141–160.

Kott, "An Introduction." In A. Kott (Ed.), Battle of Cognition. NY: Greenwood, 2008), pp. 1–9

Larsen, E. H. "A New View of C2 Systems." Marine Corps Gazette, 93 (2009), pp. 15–20.

Malone, D. B. Small Unit Leadership: A Commonsense approach. (Presidio Press Novat, CA. 1997).

Marshall, S. L. A. Men against Fire. (NY: William Morrow & Co. 1947).

McCoy, B. P. The Passion of Command: The Moral Imperative of Leadership. (Marine Corps Association, 2006).

McMaster, H. R. Crack in the Foundation Defense: Transformation and the Underlying Assumption of Dominant Knowledge in Future War. (Army War College, Carlisle Barracks Pa. 2003).

Moyle, K., & Webb, I. E. Leadership – More than "Just Good Leadership." (Paper presented at the Australian Computers in Education Conference, Cairns. 2006).

Ronen, Avihu. "The Four Leadership Traditions of the I.D.F." In Popper, M., Ronen, A., On Leadership. Tel Aviv: (I.D.F. Education Corps Pp. 1989), pp. 95–130. [Hebrew]

Siebold, G. L. "Core Issues and Theory in Military Sociology." Journal of Political and Military Sociology, 29, 2001), pp. 140–159.

Singer, P. W. "Tactical Generals: Leaders, Technology, and the Perils of Battlefield Micromanagement." Air & Space Power Journal (Summer 2009), 23(2).

Sinnreich, R. H. "Variables and Constants: How the Battle Command of Tomorrow Will Differ (Or Not) From Today's." In A. Kott (Ed.), Battle of Cognition. (NY: Greenwood. 2008), pp. 10–36

Sowers, T. S. "Nano-management: Technology, Monitoring and the Death of Professions." (PhD diss., London School of Economics. 2008).

Storr, J. P. A Command Philosophy for the Information Age: The Continuing Relevance of Mission Command. (Defense Studies, 3, 2003), pp. 119–129.

Weick, K. E. Sense Making in Organizations. (Thousand Oaks, CA: Sage. 1995).

Dr. Uzi Ben-Shalom is a military sociologist and applied researcher with over 20 years of experience in military research. Uzi is a retired LTC in the Israeli Ground Forces Command and an active reservist. His current area of research and teaching include unit cohesion in mixed gender units, leadership and decision making during terror attacks, and the role of internet traffic in stadium violence in sport. During his military service, he was responsible for writing the I.D.F. leadership doctrine. Uzi is the chair of the sociology and anthropology department at Ariel University and co-chair of the military and security community in the Israeli Sociological Association.

2

Mission Command
Between Theory and Practice
Dr. Uzi Ben-Shalom and Dr. Eitan Shamir

Dear Lou,
We have been attacking for 31 months with amazing success.
Our leaders in Tripoli, Rome, and perhaps Berlin will have
reservations, but I have taken the chance against all orders and
instructions, as this seemed to me a good opportunity. Without
doubt, it will bring great advantages in the future, and then
everybody will claim that they would have done the same thing in
my place. We have already reached our first objective, which we
were not supposed to reach until the end of May.[1]
Excerpt of a letter from Erwin Rommel to his wife, 3 April 1941

Introduction

This article will critically examine the mission command doctrine. It is based on the assumption that, although this approach has much value, there might be difficulties in applying it in armies around the world, including the I.D.F. Accordingly, many researchers contend that ranking this approach above all others might be deceptive.

In the present article, we will point out the problems inherent in applying mission command in various armies and in the I.D.F. In addition, we will attempt to characterize other command styles that have developed in Israel because of difficulties in attaining the mission command ideal. In our opinion, it is possible to create the conditions for following the

1   Basil Henry Liddell Hart (ed.), *The Rommel Papers*, St James, London, Collins, 1953.

principles of mission command, but this is not self-evident and demands that commanders be guided to act in the spirit of this doctrine.

What is Mission Command?

The literature dealing with commanding military operations on a tactical level frequently mention the importance of mission command.[2] It refers to a general approach to command based on the assumption that commanders at all levels are the most suitable to command at that level. In I.D.F. doctrine, it is defined as follows:

> The mission command approach epitomizes the assumption that every commander is totally suitable to carry out what is demanded of him at his level and in his field of expertise and by means of the force under his command. It follows that the appointed commander will excel in carrying out his mission better than any other commander, including himself. The appointee's role will expressed in determining the mission's purpose—the "what"—and in determining the frame-work in which the appointed commander meant to decide and act to carry out his part in achieving it. The appointed commander must afford his subordinates the largest possible space for decision-making and action and must avoid dictating the "how" (i.e. the operational method).[3]

Friction on the battlefield demands applying command principles based on understanding the mission and general directions from the higher command level. In other words, in a battle, the commander must be prepared to adhere to their mission and carry it out, even though they have no direct contact with headquarters. They must know when it is

---

2   The German term is *Auftragstaktik* and in English "mission command;" see also Christopher Bellamy, "Directive Control," in Richard Holmes (ed.), *The Oxford Companion to Military History*, NY, Oxford University Press, 2001.

3   General Headquarters, Operations Branch, *The Basic Doctrine for Command and Control*, 2006, p. 28, Section 5. [Hebrew]

preferable to not carry out the mission, or, alternately, when they should carry out a different mission, if circumstances dictate it.

The first instance of mission command in modern times may be found during the Napoleonic Wars in the absolute freedom awarded by Napoleon to his senior commanders.[4] The Prussian Army adopted that doctrine in the nineteenth century after Napoleon's victory at the Battle of Jena (1806). It settled in the German Army over the years and played a prominent role in that army in the Second World War. Military historians viewed the doctrine as a significant factor in the strengthening and tactical efficiency of the German Army.[5] Over time, mission command also was adopted in the lower ranks of the German Army, down to the lowest-ranking commander on the field.

This command doctrine also is considered desirable and valid on today's battlefield, which is characterized by advanced information systems and quasi-military operations.[6] Many armies around the world have recognized its importance, and it has gradually been adopted by Western armies. The first to do so was the United States Army in the 1980s in the framework of maneuvering doctrine.[7] The other NATO members later adopted it. It also was researched in Israel and, due to its obvious importance, was adopted into the I.D.F.'s official field forces doctrine.[8] According to the I.D.F.'s doctrinal literature, it is a desirable, natural, and essential command approach on the battlefield.

In addition to the mission command approach, I.D.F. battle doctrine also advocates "detailed command."[9] As opposed to mission command,

4   Interview with Martin Van Creveld, June 14th, 2007.

5   Martin Van Creveld, *Fighting Power: German and US Army Performance, 1939–1945*, Westport, Conn., Greenwood Press, 1982; Trevor Nevitt Dupuy, *A Genius for War: The German Army & General Staff 1807–1945*, Englewood Cliffs, Prentice-Hall, NJ, 1977.

6   Jim P. Storr, "A Command Philosophy for the Information Age: The Continuing Relevance of Mission Command," *Defense Studies*, 3, 2003, pp. 119–129.

7   Department of the Army, *Field Manual, 100-5, Operations*, Washington DC: GPO 1982, 1-2, 2-3, 2-7.

8   Chanan Shay Shwartz, "Command and Control in the Modern Military Organization," The Inter-Arm College for Command and Staff/Barak Program, (Internal I.D.F. Document), 1994. [Hebrew]

9   General Headquarters, Operations Branch, Field Forces Doctrine, 2006. [Hebrew]

detailed command enables close supervision over all ranks and every phase of carrying out an operation. The effectiveness of this doctrine is recognized in ground operations in commonplace, well-defined situations, such as concentrated and Special Forces missions, as well as in defined phases of conventional warfare, such as performing a planned aerial attack.[10] This command doctrine is especially widespread in I.D.F. missions against the Palestinian armed struggle.

Difficulties in Realizing the Mission Command Doctrine

Although many armies recognize the need for the mission command approach, one cannot ignore the considerable difficulties in applying it. These difficulties stem from a variety of sources and differ from one army to the next.

The literature enumerates various factors making it difficult to apply this approach. Some of them are general and related to the organization of the modern army and its missions and to advanced war materials, while some are specific and related to the unique structure of each army. The biggest difficulty in applying the doctrine in the world's armies is that military organizations fear awarding too much freedom of action to its commanders.

As stated above, it is assumed in the academic literature that the mission command approach is always essential, so it is important to strive toward it. However, this approach is not always received gladly by commanders, who fear that they will lose control, but it might be forced upon them because of the physical conditions and friction on the battlefield. The German commander-in-chief Helmut von Moltke (Moltke the Elder), who is identified more than any other general with this command doctrine, chose to give freedom of action to the generals under his command because he had no other choice. He was preparing to fight on two fronts—the Western and Eastern Fronts—and at distances

10   In English it is also called "Order Command," and in German, "Befehltaktik." See Bellamy, Note 2.

that no other general in his era was accustomed to.[11] Since the command authority defined for him did not enable him to directly command the forces, he realized that it was vital to bestow wide authority on his subordinates. Napoleon, who usually kept tight control over his armies, also was forced to give freedom of action to his army commanders. Since the huge popular army he created could not move efficiently as one entity, he divided it up into autonomous units, each acting independently.[12]

Commanders are wary of mission control for fear that it will result in excessive independence of subordinate officers in the field. From a study of generals identified with this command style, it appears that there was constant friction between them and their superior officers. Horatio Nelson, Heinz Guderian, Erwin Rommel, Orde Wingate, as well as such I.D.F. commanders such as Yigal Alon and Arik Sharon, all took the initiative in their missions—each in his own historical period according to the means at his disposal and his specific rank—thus challenging the echelons above them.

This being the case, the mission command approach is not always desirable in military organizations; even an army that considers it the optimum battle doctrine might have reservations about commanders carrying it out to the letter. Thus, although there are many occasions where the doctrine can be applied successfully, senior commanders do this unwillingly, having no other choice and out of the fear that they are opening Pandora's box, which they will be unable to control.

Mission command is based primarily on affording subordinates wide authority, but the process by which this occurs is not necessarily convenient for armies and commanders. One indication of this is the extended time that passed until the German high command awarded authority to junior officers and allowed them freedom of action in storming the Allied defenses in the First World War. The storm troopers

11   Rupert Smith, *The Utility of Force: The Art of War in the Modern World*, London, Allen Lane, 2005.
12   Yehuda Wallach, "Military Doctrines: Their Development in the 19th and 20th Centuries," Tel Aviv, Ma'arachot, 1977.

were the military mechanism that succeeded in breaking through those lines, and their success is based on the freedom given to the NCOs and junior officers leading them.[13]

Similar to other armies, the I.D.F. has given broad responsibilities to commanders, especially in situations where there was a threat to Israel's existence. Indeed, as long as Israel is threatened with extinction—a situation that characterized the first twenty-five years of the state's existence—many interpreters wrote that the I.D.F. High Command took risks and afforded its commanders extensive autonomy of action.[14]

Carrying Out Non-Military Missions

Military organizations around the world have experienced over the years a wide range of missions that do not involve actual fighting. These missions include policing, keeping the peace, and humanitarian aid. Although the armies emphasize that adherence to the mission is the most important battle doctrine, there is extensive literature dealing with the need to accustom commanders to exercise self-control and act sensitively. The relatively new image of the army as a body designed to keep the peace is diametrically opposed to the traditional image of the army as a fighting organization;[15] today, emphasis is placed on such values as empathy, tolerance, and negotiation rather than on qualities such as aggressiveness, determination, and risk taking. This indicates a general unwillingness to undertake well-defined offensives expressing adherence to a clear mission.

The expression "strategic corporal," indicating a soldier or junior officer who is liable to cause damage on the strategic level, indicates senior

13   Gunther E. Rothenberg, *The Napoleonic Wars*, London, Cassell, 1999, pp. 216–217.
14   This was the German Army's doctrinal solution of for breaking through the Allied defenses toward the end of the First World War. It was actually a fighting technique based on performing frontal attacks by several small, well-trained, well-equipped forces acting independently according to conditions in the field; see Bruce Gudmundsson, *Stormtroop Tactics: Innovation in the German Army, 1914–1918*, New York, Praeger, 1989.
15   Samuel Lyman Atwood Marshall, *Swift Sword: The Historical Record of Israel's Victory*, American Heritage Press, New York, 1967.

officers' fear of the lower ranks' taking the initiative.[16] This fear stems from understanding the need to regulate violent activities in a limited confrontation and recognizing the importance of the humanitarian dimension of the military mission. Simultaneously, there is fear of public and political criticism that may be aroused by peacekeeping missions, especially controversial ones.[17] These factors ultimately result in increased supervision by commanders over their subordinates and in greater control over their activities.

Command and Control Technology–An Illusion of Control

The development of C2 and communications technology over the past few decades has resulted in an additional difficulty for those wishing to apply the mission control approach. These advanced means of control often provide an illusory sense of control that circumvents the principle of adherence to the mission.[18] The copious exact information and close supervision characteristic of mechanized systems make it possible for senior commanders to be constantly involved in what junior commanders are doing. For example, advanced signals equipment enables the central control body to intervene in junior officers' conversations and influence their decision-making process at all levels.

With the appearance of the first C2 systems in the field brigades,[19] this ability was widely discussed in the I.D.F. and again became a topic of discussion after the Second Lebanon War.[20] Research articles have

16   Marina Nuciari, "Models and Explanations for Military Organizations: An Updated Reconsideration," in Giuseppe Caforio (Ed.), *Handbook of the Sociology of the Military*, NY, Kluwer Academic/Plenum Publishers, 2003.

17   Charles C. Krulak, "The Strategic Corporal: Leadership in the Three-Block War," *Marine Corps Gazette, 83*, 1999, pp. 18–22.

18   Eitan Shamir, "Support Operations and the 'Strategic Corporal': Implications for Military Organization and Culture," in Kobi Michael, Eyal Ben-Ari, and David Kellen (Eds.), *The Transformation of the World of Warfare and Peace Support Operations*, Westport, CT, Praeger Security International, (in press).

19   Meir Finkel, "Striving for Certainty on the Battlefield and the Inherent Dangers" *Ma'arachot*, December 2005, pp. 403-404. [Hebrew]

20   T. Heyman, "Activating the Territorial Brigade in the Hunting Era." Lecture at the second annual conference of the Institute for Tactical Analysis and Activating forces, 2005.

generally concluded that these systems were incapable of replacing command in the field. The general claim is that they are no more than a helping tool in the craft of command and control. However, in reality, they have severely curbed junior commanders' initiative and have resulted in a fighting style that is closely supervised by the higher echelons. There are those who claim that digital systems will cause mission command to disappear. Some of them are convinced that, now that commanders behind the lines can see every corner of the battlefield in real time, it may be said that the human tendency toward control and a central authority has increased with the introduction of digital C2 systems.[21]

Supervision, Control, and Bureaucracy

Various researchers and military personnel have complained in recent years that it is difficult to apply mission control in the U.S. Army,[22] which tends to act according to permanent and strict rules, thus making it difficult to apply that doctrine.[23] Other theorists have claimed that the U.S. Army has had difficulty in understanding the essence of mission control and the tremendous impact of that approach on commanders' organizational culture and behavioral patterns. It has rather focused on techniques and conventions that it learned from the German Army.[24] Furthermore, American military culture has been influenced by external approaches to management and organization that have had a strong impact on routine and military operations. For example, applying these techniques during the Vietnam War made it very difficult to create effective leadership on the battlefield.[25] This culture has become firmly rooted in the U.S. Army, despite comprehensive reforms after that war.

21   Moshe Shamir, "The War against Hezbollah: Matters of Command and Control." In Zvi Ofer (Ed.) *The Second Lebanon War—Insights into Practice*, Ministry of Defense Publications, 2008, pp. 247–291. [Hebrew]

22   Robert L. Bateman, "Force XXI and the Death of *Auftragstaktik*," *Armor, 105*, 1996, pp. 13–16.

23   Aylwyn N. Foster, "Changing the Army for Counterinsurgency Operations," *Military Review* 85, 2005, pp. 2–14.

24   Interview with Shay Hanan, June 2007.

25   Daniel J. Hughes, "Abuses of German Military History," *Military Review* 66, 1986, pp. 66—76.

Difficulties in Applying the Doctrine

The mission command approach does not only affect actual behavior in battle; it also is chiefly the result of lengthy organizational and training processes that make it possible. Martin Van Creveld, who analyzed the manner in which officer's training has developed in military history, indicated that the success or failure of officer's training might be traced back to the Prussian-German tradition.[26] Van Creveld determined that the Prussian Academy for Officer's Training was the best in the world, since it focused on all the essential processes of training battle commanders and constantly focused on the practical aspects of combat in the military profession.[27] In other words, the Prussian Academy provided basic training for senior officers and created a common relevant language expressed in a consensual battle doctrine. Mission control was the natural approach to be adopted by German officers because they trained for it.

This attitude is somewhat similar to military discipline but demands a higher level of awareness. The purpose of military discipline is to direct soldiers to act in a manner that they would not have chosen if they had not understood beforehand that it was the right thing to do.[28] Similarly, the mission command approach is the result of advanced education that prepares commanders for thought and analysis similar to problem-solving skills; it is highly likely that commanders who have undergone such training will choose to attain their objectives in this manner, although their way of doing so will differ from one commander to another. Senior officers can relax, as they know that their subordinates will act in a way that they themselves would choose, even if their conclusions are different. This stems from the fact that the lower ranks are closer to events and aware of developments on the battlefield; thus, they possess different information.

26   Richard A. Gabriel and Paul L. Savage, "The Command Crisis," *Ma'arachot*, Tel-Aviv, 1981.

27   Donald E. Vandergriff, *The Path to Victory: America's Army and the Revolution in Human Affairs*, Novato, CA, Presidio Press, 2002.

28   Martin Van Creveld, *The Training of Officers: From Military Professionalism to Irrelevance*, New York, Free Press, 1990.

Therefore, it is possible to state that one of the characteristics of mission command is an educational, professional, and intellectual infrastructure,[29] making it possible for the commander to both see their mission in the framework of the general picture viewed by their superiors and to deeply understand their commanders' intentions and the general battle situation. Such a commander will know how to use their resources correctly, even in situations where the battle situation is in flux. Thus, mission command does not only involve seeking contact with the enemy and demonstrating courage and initiative. All of these are generally true of military maneuvering in the field, but do not necessarily characterize this approach. This command style is based on insights only a deep understanding of the war phenomenon and extensive military education can provide.[30]

In the I.D.F., training in this direction is relatively limited and chiefly based on the experience accrued by officers in the course of their compulsory service. In other words, the I.D.F. officer develops primarily due to the wide experience they acquire and their personal abilities. The investment in theoretical military training is much more limited. The number of officers who receive it is relatively small, and the time devoted to it is minimal.[31]

It must be emphasized that, in such conditions, there is a good chance that commanders might wish to apply mission command, but their actions will be determined by specific battle conditions and the nature of the mission assigned to them, leading to the creation of local and one-time operational procedures. In such a situation, the connection between activities in the field and doctrine is even more uncertain. This tendency has strengthened even more in recent years, in which the I.D.F. has functioned in a military reality different to what it was accustomed to in the past—a reality in which it has been required to perform policing missions and operations against popular uprisings and sabotage.

29   Archibald Wavell, "The Good Soldier," *Ma'arachot*, Tel-Aviv, 1952. [Hebrew]
30   Van Creveld, see Note 28.
31   Interview with Hanan Shay, see Note 24.

Trust, Support, and Organizational Culture

> Up to a certain phase you search for the enemy 180 degrees in front of you and [assume that everything] behind you is safe. At a certain phase I understood that I had to be aware [what was happening in] the full 360 degrees. When they asked me why I walked around wearing high boots at Central I.D.F. Headquarters, I answered that it was because of the snakes.
>
> Chief of the General Staff (Ret.) Moshe (Bogi) Ya'alon in an interview close to his retirement from the I.D.F.

In order to apply mission command in wartime, trust is necessary among all ranks. The need for mutual trust stems from the physical distance between commanders and their superior officers. This distance demands a command style that is not based on external command and control.[32] Thus, all the military academies, without exception, emphasize the value of loyalty and internal discipline. Reliability (a kind of internal discipline) is stressed as a central value at the I.D.F. central training base.

For the last three decades, questions have arisen regarding the degree to which the I.D.F. is capable of creating trust among the various ranks against the background of organizational culture in ordinary times and routine security measures.[33] This culture is largely based on mistrust and alienation among senior commanders and accompanied by political-organizational struggles among units and corps.[34] In such a situation of internal mistrust, it is doubtful that there will be willingness to take the kind of responsibility necessary for mission command to succeed.

32   Giora Segal, "The Art of Tactical Warfare: Command in the Chaos of Battle," *Ma'arachot*, 2004, pp. 8–21. [Hebrew]

33   Morris Janowitz, *The Professional Soldier*, NY, Free Press, 1960.

34   David Pinkas, "Difficulties in Applying Mission Command in the I.D.F.," The Inter-Arm College for Command and Staff, I.D.F. internal document, 1996; see also Emanuel Weller, *The Curse of the Broken Tools*, Tel-Aviv, Schoken, 1987.

Studying the command discourse following the Second Lebanon War reveals how difficult it is for the I.D.F. to create trust among the various ranks.[35] Following that war, it was difficult to escape from hard feelings regarding the I.D.F.'s functioning and stormy arguments among senior commanders. These struggles were partially conducted in the written media, and, in some cases, rumors were spread regarding insults, leaks, and various sundry conflicts taking place in the army. This situation is far from the one prescribed by the mission command doctrine, whose central tenet is support among the ranks. It is important to note that this is not a new phenomenon, as can be seen in the expression "the war of the generals" which became common parlance after the Yom Kippur War.

Public Criticism

The past twenty years have been witness to strong public criticism of the I.D.F., focusing on such matters as fatal training accidents, especially those caused by administrative errors. The results of this criticism have varied. In part, it has assisted the army in minimizing casualties as a result of security measures, but, at the same time, it has also made it difficult to promote efficient command, especially regarding taking responsibility and a willingness to take risks. Internal army investigations, including endless investigative committees and a blurring of boundaries between the legal and military spheres, have resulted in officers who avoid taking the initiative for fear of possible mishaps, as well as an increase in authentication processes before undertaking training or operations. In addition, a slow process has begun of eroding the authority of junior commanders, due, among other things, to commanders' fear of error, an administration whose main objective is to avoid training accidents and deaths in the field, legal and criminal investigations of operational mishaps, and severe public criticism of the army.[36]

35   A. Raman, "Development of Multi-Arm Thinking," Ground Forces Headquarters, Psychological Unit, I.D.F. internal document, 2007.
36   Pinkas, see Note 34.

Interim Situations–Errors in Command Styles

The official adoption of a specific doctrine does not ensure that actual behavior will be according to the style that has been chosen.[37] Command is a social activity that takes place not only in a pure doctrinal context but also includes a wide range of other components. Furthermore, conventional warfare situations are not common; thus, command is a phenomenon that occurs on a continuum of interim situations. As a result, alongside the official doctrinal approach—i.e., mission command—there are a variety of command behaviors that represent different means of adjusting to reality. Observing these behaviors makes it possible to conceptualize command approaches without canceling out a particular doctrine or detracting from it. Such conceptualization makes it possible to name various command styles in order to discuss and study them. In addition, these command styles do not necessarily stem from individual personality traits, but instead are commonly influenced by the environment and the organizational culture in which the command process takes place.

Following are five command styles that represent interim situations and reflect errors in the process of applying mission command:

Success-Dependent Command

This is a prominent command style based on commanders receiving limited support from the higher ranks. It is accompanied by a strong sense of separation among the ranks, mistrust, and a lack of mutual understanding between the commander in the field and the higher echelons. Below are two examples of success-dependent command based on a conversation with Uri Bar-Lev, the Southern Police Commissioner, who was formerly commander of a special anti-terror unit.

In a certain mission, Bar-Lev received information about a suspicious vehicle that was about to blow up near a synagogue. When he spotted a vehicle meeting the description, he immediately stormed it and caused its

37   Ibid

occupants, who were indeed terrorists, to be paralyzed. Bar-Lev described his meeting with Minister of Defense Yitzhak Rabin when he visited the unit and the wounded:

> I asked [Rabin] what would have happened if the vehicle we attacked had not been the right one and we had injured innocent citizens and paralyzed them for life. Rabin and his adviser candidly admitted that if we had failed, they would not have backed us up . . .[38]

> In one of the operations I ordered my men to open fire on a certain vehicle. . . . I approached the vehicle to identify the dead terrorists. My hair turned white when instead I discovered inside the greengrocer, his wife, and their two children. I was in trouble—that is how it is with us. When an operation ends successfully, then everything is great, we open a bottle of champagne and are showered with compliments, "You're larger than life!" If it ends differently, then we're on our own to face harsh reality.[39]

When military operations do not end successfully, alienation develops among headquarters, commanders, and the ranks involved. Brig. Gen. (Ret.) Moshe (Chico) Tamir, regarding the war against the Hezbollah political party, may find a striking and painful expression of that alienation in an important book. As a student in the "Barak" Training Program, Tamir made many comparisons between what was desirable and undesirable in that sphere and presented numerous interim command situations that were the result of various factors that have been discussed in the present article:

> Criminal cases against field officers and a lack of support on the part of the army had implications on the units' functioning. Commanders

38   Paul Johnston, "Doctrine is Not Enough: The Effect of Doctrine on the Behavior of Armies," *Parameters, 30,* 2002, pp. 30–39.
39   Interview with Commissioner Uri Bar-Lev, *Halohem* 198, p.26, 2006.

who understood that they would not receive backing . . . preferred to avoid taking the initiative and suppressed the aggressive spirit of their units in order to avoid accidents.[40]

This command style stems from close supervision over subordinate ranks, who do not rush to assume authority due to fear of criticism on the part of their superior officers or the public. This results in complicated processes of signing off on operations that enable the higher echelons to maintain control and avoid surprises, thus creating undesirable habits. In real war conditions, such processes reduce the ability of field commanders to respond immediately as required.

Command that Places Undue Emphasis on the Mission

Despite all the obstacles and difficulties, researchers around the world and in the I.D.F. determine unconditionally that the mission command approach is necessary and vital.[41] According to them, it is impossible in wartime to wait for the green light, thereby wasting precious time; thus, senior commanders must rely on their subordinates to carry out their missions independently.

In the I.D.F., as in other military organizations, strong emphasis is placed on action and a strong sense of power. In the U.S. Army, there is a strong "can do" culture—i.e., anything is possible. In the I.D.F., a similar tendency may be found and is called *bitsuism*.[42] This means stressing performance and action and being rewarded for them, as well as emphasizing accomplishments and output. However, concepts such as power and performance ability are not parallel to mission command. The main gap between them rests in the fact that the mission command doctrine expresses systematic understanding of the mission's broader context, not only the personal ability or motivation—as impressive as they might be.[43]

40   Ibid, p. 27.
41   Moshe Tamir, "War Without Insignia," Tel Aviv, *Ma'arachot*, 2005, p. 96.
42   Storr, see Note 6.
43   Ya'acov Hasdai, "Doers and Thinkers in the I.D.F.," *The Jerusalem Post Quarterly, 24*, 1982, pp. 16–18.

Studying memoirs and war diaries reveals that commanders at all levels have often demonstrated adherence to purpose and determination, assumed the initiative and taken risks. However, does this indicate that they were acting according to the mission command doctrine? It appears that very often the answer is negative. The reason for this is that mission command largely depends on quality of training and a common language that enables understanding among the various ranks. A commander who acts very independently does not necessarily act in the framework of mission command, unless their actions strengthen the entire system and do not occur in a vacuum. In other words, the commander must understand when and when not to carry out a mission or when to alter it according to their understanding of the big picture.

A command style that overly stresses the spontaneous unique mission and stems directly from immediate tactical needs can miss the general battlefield scenario. This can stem from a lack of training. As long as the I.D.F.'s senior officer rank has relied on broad, shared battle experience, a common basis has been created for efficient command in wartime. After the Yom Kippur War, there was a significant reduction in that resource, and it seems that only fundamental military training will guarantee that it will be available in the future.[44]

Command Based on a Lack of Organization

When commanders take the initiative in battle and carry out various operations in wartime, even when these result in successes and victories, they are not necessarily acting according to mission command doctrine. Often, such command decisions are not related to aggressiveness, successful leadership, or missions, but rather to the disorder and confusion that are characteristic of war. The role of military training is to create a set of concepts making it possible to interpret reality on the

44   Eitan Shamir & Sergio Catignany, "Mission Command and *Bitsuism* in the Israeli Defense Forces: Are They Complementary or Contradictory in the New Counter-Insurgency Environment?" *Dimensions of Military Leadership*, Vol. 1, 2007, pp. 185–215.

battlefield and act accordingly.[45] When commanders are in a state of embarrassment in wartime, they are indeed capable of initiating huge, impressive campaigns. However, an examination of the circumstances that brought them to do so indicates necessity stemming from confusion rather than from military training—or even loyalty to the higher ranks or a broader understanding of the entire mission.

In this context, it is important to indicate that the common parlance used by armies indicates rapid organization according to circumstances. This has great advantages in battle but also significant disadvantages. Some examples are:

"War is chaos and we're good at chaos."

"We're kings of improvisation."

"Every plan is a basis for change."

"Learning while in motion..."

"We'll flow with it."

"This is what there is, and we'll manage to win with what we've got."

The problem with phrases such as "every plan is a basis for change" is the underlying concept that a plan is not something in which it is worthwhile investing time and effort, since whatever we do, it will be changed anyway and will necessitate improvisation. However, reality indicates that the mission command approach will only become a reality on condition that there is a master plan directing actions, thus creating a common language among all commanders on all levels. A suitable plan will create a framework that enables flexibility and a basis for discussion among junior and senior commanders. Thus, a common scenario will have formed that is a condition for change and improvisation at a later stage.[46]

Command Contingent on Receiving the Go-Ahead Signal

As stated above, in addition to the mission command approach is a command style that might be call "detailed command." In this type of

45   Interview with Jim Storr, November 2006; interview with Hanan Shay, see Note 24.
46   John Keegan, "The Face of the Battle," Tel Aviv, *Ma'arachot*, 1981.

command, extensive involvement of the higher ranks in the activities of the lower ones is considered desirable and correct. In such a situation, a command approach develops that is an extreme version of detailed command—what Benny Amidror calls "command contingent on permission" or "command dependent on superiors' permission." This command process severely curtails freedom of movement and slows down an operation's rhythm due to the need to consult superiors before making any decision. It appears that advanced information systems make it possible to function according to this approach as long as they are functional. However, it is unclear how one could transition to a freer command style if other parts of the system are accustomed to working only according to certification.[47]

It must be noted that, in an early study of the I.D.F. by Edward Luttwak from the 1950s to the 1970s, he coined the phrase "optional control." At the time when he formulated that concept, mission command was not yet so central to military thinking. Luttwak was referring to a command style in the I.D.F. in which the Israeli commander does not exercise their official authority except in cases where there is fear that the system will be seriously disrupted.[48]

It appears that the control systems of the air force and the navy, which have a firm supervisory basis, tend toward the detailed command style. Nevertheless, in both of those arms, there also exists an education infrastructure enabling the transition to a much freer command style according to circumstances. In the air force, this is expressed in the ongoing argument surrounding the independence of the leading pilots and their status in battle as opposed to the "ruling" officers posted in the main air force control tower.[49] A discussion of this issue is also present in other armies. For example, the phrase "the long screwdriver" expresses

47   Interview with Hanan Shay, see Note 24.

48   Binyamin Amidror, "The Three Command Philosophies: Mission Command, Detailed Command, and Command by Permission," I.D.F. internal document (in print), 2007.

49   Edward Luttwak and Dan Horowitz, *The Israeli Army*, London, Allen Lane, 1975.

a command conflict in which, by means of C2 systems, command headquarters interferes at a distance with subordinates' decisions.[50]

Command Toward Promotion

> How do you advance? Be careful never to take responsibility and try to avoid difficult assignments . . . then you'll advance rapidly and you'll never fail.
>
> From a lecture delivered by Maj. General Yoram ("Yaya") Yair
> at the I.D.F. Military College in 2007

Command toward promotion is a type of mission command whose goal is promotion. It involves a set of activities whose objective is to make a positive impression and draw attention. That is possibly what General Yair was referring to in the above quotation. In many cases, such activities are called a "production number" and might take place both at headquarters and in the field.

Recently in the U.S., a theory of negative leadership theory has developed called "toxic leadership." This refers to a combination of negative aspects stemming from the concept of leadership, which generally is considered an essential and positive quality. One of the negative results of such leadership is egotism and preferring personal gain to the general good or to such military values as honor.[51]

In the command context, this approach expresses the commander's need to stand out, initiate, and innovate in order to be promoted. Promotion is indeed a central factor in an organization such as the I.D.F., but extreme ambition does not necessarily lead to command in the military sense of the word.

---

50   Shmuel Gordon, "Air Leadership," Tel Aviv, *Ma'arachot*, 2003.

51   Nick Justice, "Future Battle Command and Control Systems", 2002; p. 8. Accessed from http://agilecommunications.com/agile/documents/FC2S9.pdf

Mission Command is Both Desirable and Attainable

The I.D.F.'s combat doctrine views mission command as natural, necessary, and suitable for military operations. However, applying this approach is complex and not always self-evident. The I.D.F. is not the only army struggling with this problem, and the world's armies are occupied with the question of how to train and prepare commanders in such a spirit. The "lessons" learned by various armies from the Germans in the two World Wars in the field of systematic tactical warfare are clear, but their application is difficult and not necessarily feasible. Various armies are struggling to apply this approach, but each one is meeting up with its own unique difficulties. For example, in the two World Wars, the British Army developed from a small policing force to a mass army, and this growth made it very difficult to enable freedom of action, except in very special circumstances.[52] As for the U.S. Army, its tendency to perform en masse and the need for rigid officialdom has resulted in importing management styles from the business world that advocate control and centralization. In the I.D.F., the difficulty possibly lies primarily in a lack of military education and a struggle to retain autonomy and special status in Israeli society.

As a result, it appears that one may observe an interim situation of a few command styles existing simultaneously. In other words, mission command and detailed command might not necessarily be the only ones in existence. There are cases in which these approaches are combined as indicated by doctrine, for example, in aerial and ground command. Aerial command generally is detailed during the preparation and flight to the objective, but once above the target, it changes to mission command. Regarding ground command, a few styles might complement one another due to limitations imposed on the commander's ability to take risks and the need for compromise between the possible and the plausible. Such a situation is clear in the command of large, combined operations involving

52   George E. Reed, "Toxic Leadership," *Military Review, 84,* 2004, pp. 67–71.

advanced information-gathering systems. In many phases of the operation, there will be full intelligence regarding the target, but in every mission there will be a phase where things get out of control and headquarters must pass authority on to the leading commander. Such scenarios must be practiced in advance in order to avoid misunderstandings.

It appears that traditions emphasizing initiative in battle might also benefit from combined command styles. For example, all the following statements indicate the possibility of taking responsibility at the decisive moment:

"Every soldier carries a general's baton in his rucksack." Napoleon

"You were awarded your rank in order to know when not to follow instructions!" Moltke the Elder

"You are on your own and nobody knows your situation—act logically!" A well-known slogan in the Israeli Air Force.

"Adherence to the mission in light of the mission." One of the I.D.F.'s fundamental battle principles.

A Pendulum of Command Styles

As stated above, the mission command doctrine can exist in specific circumstances, but it might also be replaced by other command approaches. The I.D.F.'s activities in the Six-Day War clearly were characterized by tactical mission command, as may be attested to by any clear-eyed observers of that war. One observer presented a few of the unwritten principles of that command style:

"If you aren't given orders, you must assume what they would have been."

"When in doubt, attack immediately."

"The battle won't turn out as you expect, so improvise."[53]

Another example of applying mission command in the Six-Day War may be seen in the words of Brig. General Uzi Eilam:

53   Interview with General Rupert Smith, June 2006.

A day after the Battle of Jerusalem ended, I decided that the battalion had nothing more to do in Jerusalem . . . I loaded my battalion onto buses and went northwards . . . I approached Dado [the Northern commander] and told him: "I'm here with a paratrooper's battalion and I'm looking for a mission."[54]

Fifteen years later in the First Lebanon War, that spirit was not so evident, but it definitely made an appearance in certain battle situations and with various commanders, for example, in the 74th Battalion's battle in Sill Village:

I meet Amos Yaron on the way . . . I'm now in his sector. He calls me aside and asks me: "What do you say?" I pull out a map and show him the plan [for the attack on Sill Village]. Afterwards . . . I assume that I have received his approval and carry it out.[55]

From an analysis of the operations that performed in the Second Lebanon War, it appears that, in various cases, commanders weighed a variety of factors when deciding upon their units' missions, while fear of casualties was a central consideration in choosing a slower advancement rate. This is a very different approach from that displayed by commanders in the Six-Day War, and it reflects, among other things, processes taking place in Israeli society and a shift in the perception of imminent threats to the country's existence.[56]

Summary

Mission command is an important and prominent command doctrine, but one cannot ignore that the doctrine exists on one hand

54   Marshall, see Note 15.
55   Brig. General Uzi Eilam, Lecture at I.D.F. Staff College, June 2007.
56   General Amiram Levine, Lecture at I.D.F. Staff College, April 2007.

and the reality of organization culture on the other. From an analysis of different command styles, a clear need emerges to identify them, acknowledge their existence, and assess that some of them are ineffectual. Furthermore, an in-depth analysis should be performed of the training process undergone by I.D.F. officers and the various organizational limitations that influence the application of mission command. Such awareness is liable to engender an internal discussion that will enable the absorption of such a doctrine. In addition, smoother transitions could have been facilitated among command styles, making clearer what is needed in any particular scenario. In the final analysis, therein lies the essence of the military profession and the art of war.

Dr. Uzi Ben-Shalom is a military sociologist and applied researcher with over 20 years of experience in military research. Uzi is a retired LTC in the Israeli Ground Forces Command and an active reservist. His current areas of research and teaching include unit cohesion in mixed gender units, leadership and decision making during terror attacks, and the role of internet traffic in stadium violence in sport. During his military service, he was responsible for writing the I.D.F. leadership doctrine. Uzi is the chair of the sociology and anthropology department at Ariel University and co-chair of the military and security community in the Israeli Sociological Association.

Dr. Eitan Shamir is a senior Lecturer at the Political Science Department, Bar Ilan University and a Senior Research Associate with the Begin Sadat Center for Strategic Studies (BESA Center). Prior to his academic position, he was in charge of the National Security Doctrine Department at the Ministry of Strategic Affairs, Prime Minister Office. Before joining the Ministry, he was a Senior Fellow at the Dado Center for Interdisciplinary Military Studies (CIMS) at the IDF General Headquarters.

His research interests and publications focus on topics such as strategy, command, military innovation and reforms, and military culture. He is the author of *Transforming Command: The Pursuit of Mission Command in the U.S., British, and Israeli Armies* (Stanford UP, 2011) as well as the editor, with Prof. Beatrice Heuser, of *Insurgencies and Counterinsurgencies: National Styles and Strategic Cultures* (Cambridge UP, 2017). He has also published articles in leading journals and various book chapters. Dr. Shamir has published in leading media outlets such as *Ha'aretz* and *The Jerusalem Post* and has been interviewed in Army Radio Channel (Galtz) and Channel 2 and 1.

Dr. Shamir has been teaching in the IDF Staff & Command College, IDF Operational Art Course, and the IDF Junior Officers academic program. He holds a Ph.D. from the Department of War Studies, King's College London.

3

Fighting Terrorist Acts from the Sea

Admiral (Ret.) Ze'ev Almog

For a period of ten years (1970–1979), acts of terror from the sea continued without the I.D.F. (in general) and the Israeli Navy (in particular) succeeding in blocking or preventing them. During that period, there was a popular saying in the navy: "It is impossible to close the sea hermetically!" Although this was quite true, there was no possibility of leaving the seacoast unprotected, as it was during those years. Thus, we were determined to put a stop to this deadly threat.

Here are the main events of that period:

- On 11 June 1971, near the Bab al-Mandeb Straits, the Israeli tanker "Corel C" was on its way to Eilat. Terrorists fired nine R.P.G. missiles in its direction. The tanker filled with fuel, yet the efficient and the speedy response of the ship's crew prevented it from exploding.

- That same decade, on 5 March 1975, there was a terrorist attack on the Savoy Hotel near the Tel-Aviv shoreline.

- The deadliest attack to occur in Israel during this period took place on the Coastal Road on 11 March 1978, which "won" the title of the "bloody bus" attack, claiming the lives of 38 and wounding 71. That fatal event took place on the Sabbath, in daylight on a major highway in the heart of Israel, and continued for several hours before it ended.

- After that attack, on 21 March 1978, Israel invaded southern Lebanon and performed a widespread retaliatory action

(Operation "Litani"). However, this did not put an end to terrorist activities originating from Lebanon.

- The attack that took place on 21 April 1979, in which a father, his two daughters, and a police inspector were murdered, was the last of those attacks to date, thanks to Israel's determination to stamp out that deadly menace to its security.

The sea attacks had their own special characteristics. The terrorist organizations employed various methods, including a detachment arriving in vessels sailing directly along the coast or from mother ships out at sea. Other methods included merchant vessels or speedboats carrying rocket launchers or large anti-aircraft batteries aimed at blowing up vital coastal installations, navy vessels, or fuel tankers; attacks on frogmen on board ships or defending essential installations; and more. Up to the appearance of the suicide bombings, there had been very few attacks in the heart of Israel, and most of them had taken place in outlying areas. Attacks from the sea afforded a convenient and almost immediate approach to population concentrations; thus, almost every terrorist landing on the coast resulted in disaster.

Characteristics of Terrorist Attacks from the Sea

- Penetration from the sea allowed arrival from long distances and from all directions.
- Such attacks did not immediately reveal the country from which the terrorists came; thus, exact identification of each vessel was required before it could be destroyed.
- It was problematic to detect vessels or reveal their identity.
- The sea was consistently full of vessels of all descriptions whose movements were protected by international law; therefore, attacking ships that were not involved in terrorist activity would also expose the Israeli merchant navy to counterattack.
- About 96 percent of Israel's freight was transported by sea;

consequently, it was obvious that caution be demanded of the country and its forces.

- The propaganda value of attacks from the sea clearly served terrorist organizations, which constantly strove to arouse public opinion and gain international notoriety.

- Successful infiltrations encouraged regular enemy navies to employ such methods, especially on the eve of war.

The combat doctrine for fighting terror from the sea developed in stages. In the early 1970s, the Israeli Navy equipped itself with small patrol boats ("Dabbur") in order to perform defensive patrols along the Mediterranean coast and the Sinai harbors and in order to economize on missile boats' expensive sea hours, preventing wear and tear and reserving them as decisive weapons. In the course of the 1970s, especially toward the end of that decade, there had been routine, almost permanent, missile boat activity far from the coast which aimed at shooting down terrorist targets. Although this activity was very costly, it did not afford any protection against terrorist attacks. In March 1978, after the attack on the coastal road, two steps were taken, the first of which was to construct fifty-eight permanent lookout posts. This was on the initiative of the head of the Operational Branch at the time, Lt. General Rafael "Raful" Eitan. The lookout posts spread out along the Mediterranean coastline and were supported by ground patrols on foot and in vehicles. This was a typical land defense response—which was very expensive but whose effectiveness was questionable, based as it was on emotion rather than logic. It did indeed prove ineffectual. These high expenses and a helplessness to solve the problem was displayed by the I.D.F. and its navy during the 1970s.

The step described above did not provide an efficient solution to the problem, as it consisted of a large number of units spread along the Israeli seacoast—whose main purpose was vacationing and pleasure—transforming it into a kind of fortress. Even worse was that the field of vision of the lookout posts was limited, amounting at night to no more than a few meters.

Upon my appointment as commander of the navy, I made every effort to tear these units down. Furthermore, the decision to establish the lookout posts had expressed a blatant lack of faith in the navy's ability to protect the coastline, which was its primary objective. Only after a year in office did I manage to convince the Chief of Staff, Raful, to back down and cancel the lookout posts, claiming that they were no more than an easy target for attack. Even back then, I claimed that the terrorists must be blocked at their starting point, either on their shores or at sea, before they managed to set foot on Israeli soil. Three events aided me in convincing Raful that I was correct:

1. Terrorists crossed the Coastal Road a short time after a Northern Command patrol jeep passed by without detecting their presence. These terrorists performed the Nahariya attack of April 1979.

2. A bilateral exercise was carried out in October 1979 between the coastal defense system commanded by the air force and an attacking naval force comprising twelve missile boats, two submarines, and nine 13th Commando landing crafts. The missile boats staged rocket fire, the commando landing craft were lowered from transport vessels, and the two submarines infiltrated the entire length of the coastline, entered it undisturbed, and laid dummy explosives and ambushes at major crossroads along the coastal road. The lookout posts detected nothing and made no attempt to prevent the attack.

3. When on a patrol mission that I performed together with Chief of Staff Eitan along the coast in 1979, we arrived together at a lookout position overlooking the Ashkelon oil harbor. Eitan witnessed with his own eyes how—after removing my insignia—I snatched the personal weapon of the scruffily dressed sergeant

staffing the lookout booth and ran off with it. Raful was furious but was finally convinced.

After the Coastal Road attack, the second step taken by the navy was dispatching missile boats against any target that crossed a line 50 kilometers from the shore without identifying itself. A statistical examination revealed that the total sea hours per month of the missile boats that were sent against those unannounced vessels amounted to 1,000.

The solution to the problem was relatively simple: the head of the commercial vehicle branch of the navy's operational department announced to the agents of the various shipping companies sailing to Israel's ports that any vessel that did not report on its arrival beforehand would be detained for a security check at the entrance to the port for at least 24 hours. This was a legitimate procedure but one that would entail high expenses for the company involved. Within a few days, all ships reported as they should, and the navy—by means of that simple instruction—saved huge sums of money.

By early 1979, I had convinced Chief of Staff Rafael Eitan that it was essential to change the combat policy against sea-based terror as follows:

- We would no longer mount reactive or retaliatory actions; rather, we would unceasingly initiate and mount attacks at all times and places where terrorists that threatened our coastline were to be found.

- We would act aggressively by means of small Special Forces, especially the 13th Commandoes. In my view, these fundamental rules are universal and applicable to any type of war against terrorism, on the land as well as the sea and coastline. Although a large part of our attack initiative was directed at the coastline, it was also valid for any other type of area.

- The system of defensive patrols and operational combings performed at sea and on enemy coasts was meant to maintain contact with the arena and not leave any vacuum that could be exploited by the terrorists. The patrols would be performed nonstop by "Dabbur," supported by missile boats and submarines, but generally not by fighter planes, which would only rarely be called upon to provide support.

- We made sure to retain, at all costs, the element of surprise regarding method, location, timing, and means. Upon receiving the Chief of the General Staff's approval for the above military policies, we applied an initiated, systematic approach that simultaneously encompassed most of the navy's spheres of activity. We mounted aggressive initiatives, continued with structural changes, and gave authority to the base commanders, thereby augmenting their control over the various sectors. We effected changes in training programs; developed and upgraded war materials while ensuring that the units quickly absorbed them; and improved the quality of the ships' commanders, lengthening their term of service in order to enable them to accrue experience.

- We improved the forces' preparedness and accelerated their training and readiness for action on sea and land; above all, we inculcated standard procedures for capturing and taking control of terrorist vessels.

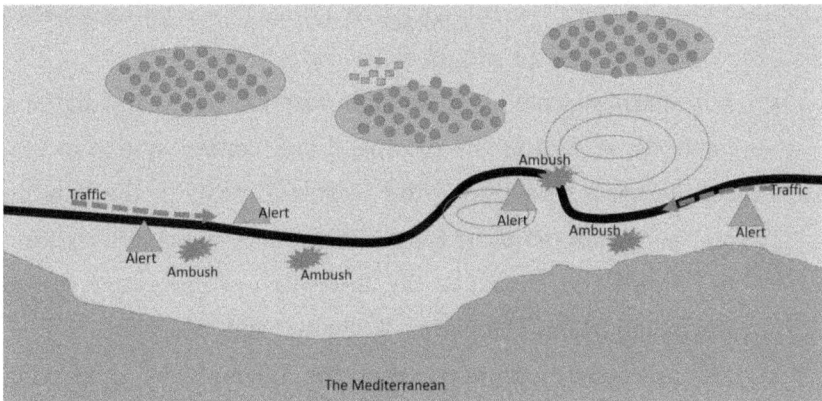

Figure 1: Ambush on the road to Beirut

The 13th Flotilla Commandoes (anti-guerrilla tactics) led missions against sea terrorism. Heavy artillery—missiles and cannons—were only employed according to need and situation to provide support or according to prior planning against particular targets. The precise professional standards demanded of the 13th Flotilla Commandoes in the numerous missions they performed considerably raised their level of ground fighting and achievements. For example, the amount of ammunition they used in each coastal ambush was reduced to an unprecedented minimum, thus increasing their swiftness of movement in the field. In addition, we planned and made sure to economize on resources, being aware that we were involved in a long, painstaking battle that must not deplete our strength. The operative and applied significance of this was our relying on focused intelligence and not on general, broad information.

The forces at sea were supplied with lists of names of terrorist vessels and their descriptions in order to be able to identify them, even in the dark. Our combat doctrine was directed at exploiting aggressive operations and operational patrols against terrorists while simultaneously preparing for total war.

Accordingly, we permanently applied a "combined naval combat doctrine," that is, close cooperation among all components of the navy's attack forces, including missile boats, submarines, and 13th Flotilla Commandoes. The results of this approach ultimately was reflected in the First Lebanon War, according to the following principles:

- We improved combat techniques, war materials, and fighting methods, as well as the command and control system at sea, on the coast, and among the command posts in the fighting arenas. We upgraded the 13th Commandoes' vessels and fighting techniques. We expedited the absorption of innovative vessels (including "Ram," "Dolphin," "Zaharon," and "Muly").
- We promoted long-range shooting methods by employing "Hilton" missiles fired from missile boats; directing artillery

fired from missile boats to shore by means of submarines or 13th Commando "Swallows," with long-distance detonation of explosives placed on the coast; and operating attack helicopters and "Hohiyot" firing mortars or sniper fire aimed at concentrations of terrorists by swimming units.

- We carefully avoided harming civilians and innocent bystanders and painstakingly identified all targets on land, roads, or ports before destroying them. We did so under the most delicate and sensitive conditions; it was mandatory to receive verification before mining a ship or firing at a vehicle thought to be carrying terrorists.

- We acted on the basis of international law regarding interrogations, arrests, taking control of ships, bringing terrorists to trial, and confiscating equipment and vessels, all while ensuring the presence of proper legal documents and the adherence to moral and legal norms as demanded by an arena in the public domain and involving neutral parties. We behaved according to these high moral standards, among other things, to prevent our troops from flagging in their purpose and to maintain morale and a fighting spirit.

- We avoided publicizing our activities in the press to prevent negative responses in the international community and retaliation by our enemies.

From early 1979 onwards, strategies for combating sea terror was modified according to a systematic, structured, and methodical program in relation to three geographical "strips" stretching between enemy territory and our own coastline:

- Strip No. 1 included the enemy coastline used to prepare and organize for sea attacks on our shores.

- Strip No. 2 included the stretch of sea between the enemy's and our own shoreline.

- Strip No. 3 included the defense and deployment system on the Israeli coastline and its infrastructure.

Our activities in each of these three strips was directed at thinning out enemy targets at the rear in order to minimize the number of targets/threats that we would need to deal with on the stretch of sea near the Israeli shoreline.

Each Strip's Activities

Strip No. 1: The enemy coastline. Striking at the source by attacking terrorists preparing to set out on a sea attack, thus preventing them from doing so and forcing them to defend themselves. Activities in this strip mainly took the form of initiated attacks, with the following aims:

- There were attempts at intensive and continual attacks against specific locations (except for occasional breaks dictated by other vital considerations) against organizing or training areas, headquarters, arms storehouses, transport axes, embarkation ports, points of arrival from the sea—such as harbors—and vessels in service of the terrorists, including those being used for arms shipments by sea.

- Emphasis was placed on continuity, variety, and surprise in any location where a potential threat was detected and along the entire Lebanese coast (or outside of it). It stressed that operations were not to be limited to points close to the Israeli border or from which the terrorists had set out on their most recent attack; this was done in order to maintain the element of surprise and strengthen deterrent value.

- Fighting in Strip No. 1 was spread over three periods, while the nature of the activities changed according to an obligatory list of priorities and depended on circumstances and military and political developments in the region. The fighting arenas according to situation, time, and place were as follows:

  1. January 1979–April 1981: We blocked terrorist attacks and mounted counterattacks, especially in the sector between Rosh Hanikra and Beirut.

2. May 1981–December 1982: We exploited contact with the battle arena because of intensive preparations for the First Lebanon War (in the course of that operation and afterwards). These actions were taken in order to maintain attacks against terrorists on the coast between southern Lebanon and Beirut and block their access to the sea. Thus, we managed to push them northwards toward the Tripoli area in northern Lebanon.

3. January 1983–January 1985: We built on the results of the fighting against terrorists in general and along the Tripoli coastline in particular until they also vacated that area. In this manner, we damaged the terrorists' renewed efforts to restore their sea power in the Beirut area. In addition, we damaged or captured vessels setting out from their countries of exile in an attempt to infiltrate war materials and terrorists to our shores.

Strip No. 1's Fighting Methods

Ambushes were conducted along the Lebanese coastal highway, which were based on these unique fighting techniques and specifications:

- The ground ambushes divided into two teams: one alarming, the other shooting. This technique was aimed at ensuring correct identification and efficient targeting of suspicious vehicles and terrorists only.

- Arrival at the shoreline was performed in absolute secrecy (by means of swimming, diving, or "Hazir"). In that spirit and according to those methods, arrival by sea was planned and carried out by both the navy vessels and the 13th Commandoes. This technique made it possible to land or to approach the target at almost any time and place and in all maritime or landing conditions, such as rocks, breakwaters, etc. Thus, the 13th Commandoes' special advantages were exploited to the full. In this way, the terrorists' night activities were minimized until they practically ceased to exist.

There were a range of targets and terrorist forces from the sea, with examples of activities performed by us in various combinations:

- In order to destroy the Katyusha battery that fired every evening on Nahariya, a submarine neared the coastline at dusk while a missile boat was simultaneously positioned a few kilometers from the shore (out of earshot). Until then, attempts to employ ground artillery, planes, and fighter planes had not succeeded in silencing the battery due to its vanishing into the darkness immediately after the launching. Firing from the missile boat was performed with the guidance of the submarine (whose narrow, undetectable nose pointed toward the shore). The missile boat was secretly anchored a fair distance from the coast. Within thirty seconds of the Katyusha's launching, the missile boat fired its guns, guided precisely by the wireless signals transmitted by the submarine regarding the Katyusha's exact firing position on the shore; the Katyusha was destroyed with the first salvo. The signal transmitted to the missile boat relied on the flash of fire observed from the submarine's bridge and relative to an orientation point on the coast that had been prepared beforehand.

- A submarine patrolling by periscope in daylight directed the fire of a missile boat located far out at sea in order to destroy forces, vehicles, installations, gun batteries, and bridges. This method was widely employed during the First Lebanon War.

- A "Snunit" neared the shore under cover of darkness and illuminated a terrorist headquarters building; a missile boat anchored at sea launched a missile controlled by the direction finder on its bridge and accurately hit the illuminated structure. The launching was aimed at the signals installation of the combined terrorist headquarters on the Al-Barad Bridge, opposite radar-controlled Syrian land guns, which limited the approach of vessels the size of missile boats. There was a refugee camp at the rear of the detonated installation, and the prime minister at the time, Yitzhak Shamir, demanded my personal guarantee that the missiles would not reach the camp. I gave my word and kept it.

- Under cover of darkness, submarines affixed a leech mine to a terrorist ship located in the Tripoli harbor in Lebanon. At first light, missile boats that were anchored outside the harbor bombed it and the war materials found in the area at the time; as a result, the ship capsized. Our intention was to make an impression regarding the strength of our bombs.

- Attackers from the 13th Commandoes disembarked from missile boats under cover of darkness, sailed toward the shore on board 13th Commando transport vessels, jumped into the sea, and swam to shore, where they laid mines that detonated the next day at noon. These actions were done in accordance with surveillance done by means of a submarine's periscope, adjacent to an armed force traveling and marching along the coastal road.

- Two sea helicopters took off from the deck of a helicopter-carrying missile boat sailing far from the coast and fired missiles in daylight at a terrorist detachment (the 17th) north of Tripoli harbor. A second missile boat finished off the operation with gunfire. This method was chosen due to the limited possibilities of nearing the coast due to the Syrian/Palestinian artillery positioned along it.

- Sinking ships served terrorists in Lebanese harbors (Tsor, Sidon, and Tripoli) by means of frogmen who affixed leech mines to the vessels or hung explosives near breakwaters in order to arouse panic. Affixing leech mines to ships was carried out only after a supervised identification process occurred by means of reports verifications by radio contact to receive the go-ahead from navy headquarters (despite the risks involved in that procedure).

Ground attacks:

1.   The target for attack was a 13th Commando raid (Operation "High Voltage," 17–18 April 1980) on a terrorist structure on the eve of a second attack on Kibbutz Misgav Am. The building and some positions were blown up, killing fifteen terrorists.

2.   A mortar attack was made on a training base far away from the coast by means of swimming under cover from the sea (14–15 September, 1983).

Strip No. 2's Fighting Methods

Strip No. 2's attacks focused on offshore activities aimed at deliberate harassment, taking control of suspicious vessels at sea, and stopping their progress.

In the first phase, we distanced international shipping from the sea routes opposite the eastern Mediterranean coastline to those farther out at sea. Ships that were identified were interrogated; suspicious ones were stopped and brought to Israel.

Year	Identified/interrogated ships		Taken control of ships		Stopped ships brought to Israeli territorial waters
	Mediterranean Sea	Red Sea	Mediterranean Sea	Red Sea	
1978	67	26	22	4	3
1979	354	212	24	none	10
1980	51	55	4	none	none
1981	8	1	none	none	5
1982	218	none	13	none	1
1983	22	none	none	none	4
1984 (July)	2	1	none	none	3

Table 1: Israeli Navy ships monitoring activities, 1978–1984

In 1979 alone, from the time of adopting the new policy of interrogating, stopping, and taking control of ships, we took over twenty-four suspicious vessels, while ten of them were brought to Israel and stopped. On those vessels, terrorists were arrested and war materials, explosives, and ammunition confiscated. The results of this strategy constituted a big step forward compared to 1978.

The interrogation and identification procedures taking place in 1979 totaled 354 in the Mediterranean and 212 in the Red Sea. They aimed at distancing the bulk of sea traffic from our shores and represented a considerable increase compared to 1978 and a drastic reduction in activity the following year; in 1980, there were only fifty-one interrogations in the Mediterranean and fifty-five in the Red Sea. This development verified and demonstrated that the policy adopted by us was bearing fruit. Indeed, the bulk of international shipping kept its distance, making our identification and patrolling activities much simpler.

The renewed increase in interrogations and examinations that took place in 1982—218 interrogations and thirteen cases of taking control—was due to our intensified operations during the Lebanon War and the blockade that we imposed on the Lebanese coast.

The bothersome interrogations were mainly carried out by missile boats. They were applicable to any ship sailing along those sea routes due to a (legitimate) intention to cause delays and discomfort, but without stopping the ship or taking control of it. Experience taught us that innocent passenger and cargo ships preferred to avoid such delays; therefore, they distanced themselves westwards on their own initiative, 80-to-100 sea miles from the shore, thus saving themselves any trouble. This considerably minimized the bulk of traffic and the number of ships passing 50-to-80 sea miles from shore. Consequently, this made the ability to identify the remaining vessels much easier for us, especially in regard to those vessels approaching our coast capable of transporting small enemy craft that could easily reach our shores.

We took control of suspect vessels and stopped their progress according to international law and based on defense regulations. This was according to intelligence information and behavior or ships displaying suspicious features, such as rubber dinghies on their decks.

A very different boarding procedure was implemented, in which the process of taking control transferred to areas opposite our coastline. Boarding was carried out in a professional, careful, and confident manner by a trained 13th Commandoes force undercover with the surveillance of a helicopter circling above. Terrorists and war materials that were captured were dealt with by military courts; cargo and vessels were confiscated and terrorists were arrested according to court order. Terrorists on these ships (some of them senior officers) were arrested, and large amounts of fighting equipment—especially tanks, land guns, light weapons, explosives, ammunition, and signals equipment—were confiscated.

We consistently avoided publicizing these operations—against the wishes of Yehoshua Sagi and Ehud Barak, the heads of intelligence at the time, who considered that publication would constitute a kind of deterrent. We contended that our doing so would cause more harm than good, as it might result in Israeli ships being captured without legal justification. In any case, the terrorist organizations awaiting the ships were well aware of what was delaying their arrival, so there was no need to "wave it in their faces." The two chiefs of staff at that period, Lt. General Rafael Eitan and LTG Moshe Levi, accepted our position on this matter.

At the second phase, permanent or routine patrols ("Raz" and "Bar") performed at sea by missile boats were canceled and replaced by interception flights based on concrete knowledge. The reasons for this were as follows:

- There was little likelihood of encountering a terrorist vessel along the patrol route stretching 20-to-30 miles from the shore, as there was no assurance that a patrol vessel nearing the end of its route would detect an enemy vessel at its far end.

- Intercepting suspicious vessels based on concrete information and not on general, unfounded warnings proved effective on numerous occasions. Thus, thousands of sea patrol hours were saved, as well as wear and tear on vessels and crews.
- The interceptions were carried out by ships already at sea for training or for operational patrols or other missions (not necessarily originating from their bases). These ships (including missile boats, light "Dabbur" patrol boats, training ships, submarines, and tank landing craft) all were supplied with lists and photographs of ships in service of the terrorists and were trained to spot them (even in the dark), thus saving many sea hours.
- Cargo vessels waiting to enter Israeli ports that did not submit reports according to shipping regulations but entered the fifty-km zone from the coast were informed that they would be detained for security checks lasting 24 hours. They immediately responded, saving us 1,000 sea hours per month.

Strip No. 3 Fighting Methods

Extensive reorganization was done throughout the command system, defense of the Israeli coastline, and its infrastructures. The command structure, responsibilities, and authority of the naval bases were transformed to be sector-bases as follows:

- The base commanders were made responsible for whatever took place in their sector, stretching to a depth of 30 miles from the shoreline.
- The base and sector commanders were given authority (similar to that of ground sector commanders) over all the forces stationed at their bases or moving through their sectors. It followed that this included all of the vessels and installations in the sector/base, including missile boats and submarines anchored at the Haifa base that in the past had not been subordinated to the

base commander but were logistic only. Now, all movements of vessels in the base/sector were directly under the commander's authority and were their responsibility. The base commanders were fully responsible for preparing all the vessels in their bases/sectors for war and operations. The powers granted them (that were based on naval commands) motivated them to act to exploit their abilities and experience in order to prevent and confound any infiltration into their territory. The navy no longer depended on long-distance control.

- The command of the light "Dabbur" patrol boats, which carried out most of the field security missions near the shore, was handed over to officers who henceforth acquired wider, more fundamental command training and experience than in the past. The authority to command small, independent vessels, such as "Dabbur" and tank landing crafts, was now bestowed two years after completing the officer's training course (instead of only half a year, as before). In addition, young officers now received command only after completing a comprehensive eight-month commander's course. Thus, they were well prepared for independent command and trained to engage enemy forces and international parties discreetly and responsibly.

- Development began of a coastal detection system based on radar that would take five years to complete. The radar (developed by ALTA) was absorbed by the navy on the sixth year of my service as naval commander and, as expected, caused a revolution in the field of coastal detection and control. It resulted in the following:

  ‣ It was now possible to detect and discern targets, even small ones, much more successfully in all weather and sea conditions. For example, the bus attack on the coastal road could have been averted, as it took place when weather conditions were too stormy to detect the rubber dinghy that transported the terrorists to shore; the radar in operation at the time was incapable of providing such detection.

> The gap narrowed for detecting objects in dead ranges close to shore. In the Nahariya attack, the terrorists had exploited this gap. The new "Victor" radar was capable of detecting a sea motorcycle (a small target) operated by a terrorist near Rosh Hanikra, while off the coast of Eilat it even managed to detect swimmers in the Red Sea.

> It was possible to reduce the number of radar stations along the coast from eighteen to eight and equip them with one system that was relatively cheap and easy to learn to use, operate, and maintain.

Conclusion

- Between 1970 and 1993, during its war against terror, the Israeli Navy damaged 101 ships and crafts that were involved in terror from the sea. On fifty-eight different occasions, according to the following division: twenty cargo ships, thirty-six fishing boats, twenty-six motorboats, seventeen rubber dinghies, one tugboat, and one sea motorcycle. Out of those vessels, seventy-four were sunk.

- During the six years of fighting against sea terror (1979–1984), even during the First Lebanon War, there was no loss of life among navy personnel.

- During those years, the navy performed 400 percent more operational sallies than in 1973, which was a war year.

- In 1981, the chief of the general staff awarded Medals of Honor to the 13th Commandoes for their efforts and successes in blocking terrorism during those years. Prime Minister Menachem Begin, acting as defense minister, granted the navy the sum of three million dollars—beyond its usual budget— as a gesture of esteem, which the navy was authorized to invest in any project it chose. The funds were invested in port security.

Offensive Action of the Israeli Navy While Combating Terror from the Sea

In the course of fighting terrorism, we felt a serious lack of fast patrol vessels, both for field security and attack missions comprising attack, identification, and taking control. What was required was a vessel that could arrive quickly—faster than 50 knots—to any suspicious vessel, even in a choppy sea and opposite various types of coasts—including shallow water, characterized by sandbars and obstacles—in order to identify it before it escaped. We required a patrol boat that would provide the crew with stability, a dry deck, maneuverability, and sustenance; that would enable a good, lively performance; prevent seasickness and jolts; and be maintained without too much expense or complication. Such a patrol vessel was developed on my initiative and under my supervision after my release from active service, while I was manager of Israel Shipyards Ltd. It was given the name *Shaldag* (kingfisher).

The struggle with terrorism from the sea—which was almost entirely performed according to the initiative and decisions of the navy itself that was thereby presented with a daily challenge—was like a sea voyage on calm waters compared to the campaign I was forced to undertake regarding the design, planning, and development of future war materials for the navy (including *Shaldag*).

Figure 2: *Shaldag* patrol boat initial prototype

A True Change in Direction is Required

The terrorist attack that took place in Nahariya on 22 April 1979, in which four residents were murdered, was the last terrorist attack from the sea at the end of a blood-soaked decade. From then on, there were no more successful attempts to reach the Israeli coast from the sea. The coastal sector was the only one in Israel in which terrorist attacks were completely stopped, and this unique achievement has remained in place for the past thirty-six years. At the same time, extensive I.D.F. operations against terrorists—Operation "Litani," followed by the First Lebanon War—brought the army to occupy the security zone in southern Lebanon for eighteen years, from June 1982 to the retreat to the international border in 2000. Other operations followed, including Operation "Defensive Shield," Operation "Determined Path," Operation "Iron Law," and Operation "Protective Edge." While multi-dimensional, these operations did not produce results equal to those achieved on the sea and the coastal sector.

It cannot be said that sea warfare is fundamentally different from, or less complex than, ground warfare. In fact, it was as demanding—or even more demanding—than ground fighting. Since the fighting arena involved multiple targets on land and sea, the need to identify them positively in order to avoid harming the innocent was critical, and dealing with difficult physical conditions and a constant international presence.

It is obvious that the fundamentals and rules for combating terror are universal. Consequently, we should consider this question: Is it not time to change direction by fighting terror by means of special units and small-scale fighting and abandoning the use of large-scale powerful formations that do not achieve their objectives and strike a mighty blow but miss the target? In most cases, these ground operations were forced to a halt due to international pressure. In other words, instead of mounting huge campaigns—involving fighter planes, artillery, tanks, missile boats, bombs, and widespread bombardment—that injure innocent bystanders and do not lead to long-term results, the focus should be on small-scale special units. It is necessary, then, to examine and apply the lessons and strategies that guided the navy and resulted in its decisive victory in this field.

Admiral Ze'ev Almog joined the Nahal Infantry Brigade in 1952 and, in 1954, volunteered for Flotilla 13 where he was qualified as a combatant. In December 1957, he graduated the Israeli Naval Academy and subsequently held several command positions in Flotilla 13 and in the Torpedo Boat Unit. In 1965, he received his B.A. from the Hebrew University of Jerusalem. In 1968–1971 (the War of Attrition), he commanded Flotilla 13. In 1972–1974 (the Yom Kippur War), he was commander of the Red Sea Arena. In 1975, he graduated from the US Naval War College and received an M.A. in Strategic Studies and Administration. In 1977–1978, he was a member of the founding team of the National Security College, where he also was an instructor. In 1979–1985, he served as Commander of the Israeli Navy and played a major role in shaping Israel's modern Naval Force. In 1986–1995, he rehabilitated the Israel Shipyards, which had been on the verge of bankruptcy and in temporary liquidation, by privatizing it and restoring its profitability.

Admiral Almog has published several books, including *Bats in the Red Sea* (2007); *Flotilla 13* (2011, published by the U.S. Naval Press); and *Commander Flotilla 13—The Sails of My Life; 13 Flotilla* (2017, in Polish).

4

Mission Command and Intelligence

A Built-In Paradox

Colonel (Res.) Shay Shabtai

Abstract

The connection between intelligence work and characteristics of mission command creates a built-in paradox. On the one hand, the precise data obtained by the intelligence services regarding the enemy allow fighting forces to conduct battles in a more independent manner. Freedom of action in the spirit of mission command is also expressed through intelligence work itself, thereby enabling creative freedom to its personnel in order to obtain the required data, analyze it, and act accordingly. On the other hand, intelligence organizations serve decision makers and senior officers and can supply them with much information regarding the enemy—down to the micro-tactical details—enabling them to centralize their command. These two extremes are further demarcated by technology and the evolving human factor. The great challenge facing those considering the future of warfare is to redesign the application of mission command in an intelligence directorate that has become such an essential support in light of the information revolution and the growing importance of intelligence to command and operations.

Background: Mission Command

In *Transforming Command*, Dr. Eitan Shamir has characterized mission command as the connection between the standardization of outcomes and the standardization of abilities. The fighting force's ability to achieve what is demanded of it is based on the high professionalism of the individual enabling an appropriate understanding of the necessary tactical outcomes (the mission) in order to realize strategic goals (the objective) and exercise independent thought and action in the spirit of mutual understanding in face of challenges and uncertainty on the battlefield. In such a situation, the operational system's efforts to achieve its objectives are greater than the sum total of its actions. This constitutes the highest realization of the military organization's potential.

Mission command demands delegation of powers on the part of commanders: giving freedom of action to their subordinates; empowering them by means of guidance, mutual trust based on professional training, and constant mutual communication; and a learning process and toleration of mistakes, thereby allowing their taking initiative and having the authority and responsibility to carry it out. All of these qualities contrast strongly with centralized command, in which the senior commander controls information down to the smallest detail and consequently exerts micro-command on each activity of the lower echelons.

Mission Command and Intelligence: A Paradox

My students and I have defined intelligence as a recognized professional area, which encompasses processes and organizations providing the required information regarding the abilities, intentions, and activities of foreign or hostile elements as a means of facilitating decisions and actions and, in addition, of carrying out clandestine activities.

The definition of intelligence and the characteristics of mission command create an inherent paradox. On the one hand, the precise data obtained by intelligence regarding the enemy, their capabilities, and their location makes it possible for the combat forces to carry out fighting in a more independent and efficient manner. Independence of action in the spirit of mission command also comes to expression in intelligence work itself, enabling creative freedom to its personnel—who are generally intelligent and daring—to achieve the necessary data, analyze it, and act accordingly.

On the other hand, intelligence organizations are active on the national level, the level of the general staff, and the headquarters of the campaign and the tactical level (command, corps, division, and brigade). They are in contact with decision makers and senior commanders to whom they are able to supply large amounts of information about the enemy, down to the smallest detail, in a manner that can facilitate centralized command.

In this chapter, I will examine the built-in paradox between intelligence and mission command through examples drawn from the Israeli intelligence experience. I will do so on the following three levels:

- Mission command in intelligence: Granting authority and enabling creativity in intelligence work.
- Mission command of intelligence: Freedom of action allowed to intelligence operatives in the field during their missions.
- Mission command and intelligence: The function of intelligence as a support to mission command in the main operational echelon.

Mission Command in Intelligence

The main role of intelligence is to supply data regarding the capabilities, intentions, and possible activities of foreign or hostile elements in order to facilitate reaching conclusions and taking action. I will analyze the subject of mission command in this context in three dimensions.

The first dimension is coping with three types of challenges: the secret, the enigma, and the mystery. The secret is a question that has an obvious and generally accurate answer (i.e., the enemy's capabilities or what activities they are planning). What is required in order to decode the secret is tapping intelligence sources that penetrate to the heart of the enemy's system and unveil hidden data and intelligence analysis that will construct the copious data into an assessment of what the enemy is hiding from us. To do this, it is necessary to set up an intelligence "factory" centered on defining the questions and delegating resources to collection, processing, and analyzing in order to answer them. Mission command is limited here to types of work methods.

The enigma is a question that does not have a definite answer, since it involves predicting future developments (i.e., what the enemy is planning to do). Nevertheless, it is possible to gather and analyze data that will reduce the number of possible answers, such as the enemy's own statements regarding its future activities and its possible timetable. What is necessary in order to solve the enigma is a combination of intelligence agencies concentrating on creative, unfettered thinking and a dialogue with the intelligence "client" in order to clarify the complexity of the matter at hand and present the various possibilities that must be taken into consideration.

The mystery is a broader question related to deep-seated human processes or events that are currently developing (e.g., whether a revolution will break out). The relevant data that can be collected and analyzed is of limited value. Thus, an open dialogue is required between intelligence and the client in which various scenarios are discussed—as well as their repercussions for decision-making.

The table below presents the relationship between the level of mission command—centralized and limited or decentralized and extensive—afforded to the intelligence directorate in the face of the three challenges described above.

	Secret	Enigma	Mystery
Defining the problem and ways of solving it	Centralized	Centralized	Decentralized
Constructing the intelligence's answer to the question	Centralized	Decentralized	Decentralized
Dialogue with the client Centralized	Centralized	Centralized	Centralized

Table 2: The relationship between levels of mission command

It is possible to indicate two opposing developmental directions. On the one hand, the current nature of the regional and global environment intensifies the mystery, thus demanding a more decentralized approach. Conversely, as I will demonstrate below, on the background of growing uncertainty, intelligence focuses on an easier quest to solve the secret, which is centralized in nature. In any event, the dialogue between intelligence and the client—particularly decision makers—was and will remain centralized.

The second dimension, which has accompanied the intelligence community for many years, especially after the failure of early warning before the 1973 Yom Kippur War, is pluralism in analysis, the aim of which is to open the discussion to multiple assessments and minimize the danger of systemic failure.

The head of the I.D.F. Intelligence Directorate in 1973, General (Res.) Eli Zeira, was asked about diverse opinions on the eve of the war and replied: "That is exactly what bothers me. Of all the analysts who appeared before me, why wasn't there anybody with a different opinion?" On 6 October, the day war broke out, at 0630 hours, Zeira had known of at least three officers in the production and analysis department—the head of the Jordanian Branch, the head of the Military Deployment and Terrain VISINT Analysis Branch, and the head of the Technological Branch. They had told Zeira of their assessment that war was about to break out. At 0715 hours, Zeira presented the General Staff with his assessment that it was illogical for Egypt and Syria to go to war, although signs on the ground indicated that they were about to do so, and he made this assessment without taking into consideration the opinion of his senior officers. In other words, pluralism in analysis depends on the character and personal preferences of the heads of the intelligence services and often might not get through to decision makers.

In its inquiry into the 1973 War, the Agranat Commission determined that

[A]pparently there are advantages to the system of one unified assessment by the Military Intelligence Directorate at the end of each discussion, as it seemingly presents its clients with a firm basis for reaching decisions. However, this is an illusion, as fundamental differences among analysts might not come to those clients' attention.

In light of the above, the commission recommended that "it was not to be ruled out that opinions opposed to those prevalent in the intelligence directorate should be presented in written briefings and assessments. In fact, an open presentation of such differences enables readers to understand that the analysis is possibly inaccurate, enabling more caution in taking operative steps."

The Agranat Commission's recommendation to present internal differences of opinion became a rule of thumb in the intelligence services in the years following the war. Forty years after the Yom Kippur War, Brig. General Itai Broon, head of the Production and Analysis Division, wrote that "The only way to advance our understanding of reality is by means of ongoing, systematic brainstorming based on debate. Debate affords the opportunity to give warning of everything that is perceived to be unconventional, impossible or inconceivable." The basis for the ability to carry out such processes depends on "an organizational culture that encourages openness to a wide range of opinions, including unusual ones . . . an environment enabling creativity and intellectual freedom. Without this, intelligence research will find it difficult to fulfill its goals in everything concerning clarifying and comprehending reality."

At the same time, aiming at pluralism and a culture of creativity does not lessen the difficulty of providing relevant intelligence assessments to decision makers. Considerable efforts have to be made to construct and provide cognitive tools for assessment. This need has gradually increased due to considerable uncertainty about the future, uncertainty brought about by the upheavals in the Middle East starting from the end of 2010.

Itai Broon has suggested coping with this through the analysis of competing hypotheses, an idea developed by an intelligence methodologist and former CIA analyst, Richard Heuer, which is based on "raising a wide range of explanations (regarding the present) and possibilities (regarding the future)," then "collecting data actively in such a way as to refute some assessments and support others" (Brun).

A senior intelligence officer raised the idea of "a wider sum total of assumptions enabling us to reach a larger number of conclusions . . . In a process of this sum total of assumptions, we gradually create a network of assumptions that are connected by deductive links . . . Rational behavior is related to broadening this network of assumptions together with ongoing critical activity in order to maintain its coherence."

Observing mission command in light of pluralism in intelligence analysis, it is possible to identify a striving toward decentralism and freedom of action. However, this suffers from another problem: the difficulty of establishing the connection between standardization of products and standardization of qualifications. Over the years, analysis units have found it difficult to determine a standard of analysts' qualifications.

The third dimension is the role of the collection array, that is, of the language analyst listening to conversations and the SIGINT team producing them, the GEOINT analyst dealing with massive amount of imageries, the HUMINT officer examining sources and their information, and the data mining analyst going over huge stores of textual information from open and classified databases. All of these roles have both power and responsibility, as they are the sensors of the intelligence community dealing with the primary sources and are in the most intimate contact with the subjects under examination.

These officers suffered from a great sense of frustration following the '73 War, as they felt that they had provided a clear warning that had not been heeded. Brig. General Yoel Ben-Porat, commander of the Surveillance Unit of the Military Intelligence Directorate (known today as "8200"), strongly expressed these feelings in his book about the Yom Kippur war: "Our limitless ability to gather information increased by 1973 . . . but it was not enough. The fate of this information was determined by another body: the Military Intelligence Directorate's Production and Analysis Department . . . from its establishment it was responsible for numerous unjustified blunders that occurred either due to improperly using information, ignoring it entirely, or being incapable of viewing reality, either due to their ignorance of Arabs, the Arabic language, or the Islamic religion."

Ben-Porat draws the following conclusion:

I should have reduced my attentiveness to commanding the unit and concentrated on what the I.D.F. was doing, starting from October 4—the size of its forces in the Sinai and the Golan Heights—and examine if the picture presented to the General Staff corresponded with what emerged from the intelligence data. I did not do so due to the 'Trust me' system prevalent in the I.D.F., according to which it was unacceptable to 'stick one's nose in' and overstep authority, but rather to rely on the system . . . Looking back, this was unwise and totally futile.

The lesson he derived from all this was that he "should have greatly overstepped [his] authority and examined what was happening with the intelligence info [he] was supplying and how the active I.D.F. forces who received it were applying it."

Ben-Porat's conclusions became firmly rooted in "8200" in the following decades. In 2003, thirty years later, it was still prominent, as is apparent in the following:

[O]n the unit['s] internal website. In '8200' (and in the Security Services generally) a concept developed of 'the duty to issue a personal warning': if an officer or soldier sensed that the information he was providing was not being treated as it should, it was his duty to immediately inform his senior officers . . . The current unit commander has placed special emphasis on this. On the website, which is open to all members of the unit, he presents all the telephone numbers at which he can be reached (at work, at home and by cell phone) together with his e-mail address, and urges his subordinates to contact him as necessary.[1]

In the past decade, it appears that the relationship between information collection agencies and analysis units has undergone a change. The article "Intelligence 2.0: A New Approach to the Production of Intelligence" discusses this change. It recommends a transition from a cyclical system to a work method in the spirit of Web 2.0, according to which the user changes from a data client to a content creator. In this framework, communities of experts create a shared intelligence network. The network will connect the collection and analysis bodies with intelligence's clients, who will all cooperate in developing applicable data while preserving disciplinarian expertise. Such a procedure enables collection agencies—as Yoel Ben-Porat envisioned—of becoming real-time participants in the analysis and operational dialogue stemming from the data they provide.

Regarding the role of information collection in light of mission command, it appears that an inherent tension exists between the need to develop a more sophisticated collection, sorting, and analyzing system in light of the information explosion, which is centralized in nature, and the need for a creative brainstorming dialogue between intelligence and its clients in order to refine the final product. Technological innovations in the information sphere have provided them with better tools for combining the two.

1   Amos Harel, "In 8200, they are praising the conscientious objector, but criticizing his behavior." Haaretz, January 31, 2003. [Hebrew]

To sum up the analysis of the level of mission command prevalent in the intelligence, one can say that in the three dimensions that were examined—coping with challenges of assessment, pluralism in analysis, and the role of collection array—there is a blatant paradox between the need to present a clear, reliable product to clients and decision makers on various levels and the need to delegate authority and enable creativity during its preparation in order to make it the best intelligence possible, especially in areas of uncertainty and the need to predict future developments.

Mission Command of Intelligence

The intelligence directorate is also an operational body active in the clandestine field in order to access data in places that can only be penetrated by means of complex operations, including reaching the enemy's most carefully guarded locations. In addition, intelligence acts to target individuals and infrastructures in order to neutralize threats and planned attacks.

In order to carry out such missions, cutting-edge elite operational intelligence units are called into play that are at least partially cut off from the chain of command, as they must overcome unexpected developments in an operational arena fraught with dangers. Such a situation demands that mission command be applied fully.

I will give examples of such challenges in three areas: special operations, HUMINT operations, and operations in cooperation with allies. In special operations, this story is told of a special commando team from "Sayeret Matkal" (G.H.Q. Special Force), headed by a young officer, Ehud Barak (Israel's future prime minister):

After weeks of preparations and training, the force set out on its mission. Navigation was difficult after several hours of marching and crossing a water obstacle, Ehud spotted the target on the outskirts of a Syrian army camp . . . While they were skirting the Syrian post, his signal operator whispered an order by the forward command that they had run out of time and must retrace their steps. Ehud comments: 'We were about fifty meters from the camp's outward sentry post . . . I was sure that we could complete the mission and get back in time. I could not explain this over the wireless, which was an old Motorola model. The noise it made was terrible and I was afraid they would hear us. I ordered the man holding the device to shut it off and we went forward . . . That was the first time a unit performed a mission crossing the Syrian border and it won Ehud his first medal for bravery. (Kfir and Dor)

In the field of HUMINT operations, there is the well-known case of Yehuda Gil, a Mossad operative who supposedly developed a relationship with a senior Syrian official and reported false information that almost led to the outbreak of war between Syria and Israel in 1996. Ilan Mizrahi, who at the time was head of the Tzomet Directorate of the Mossad, which was responsible for recruiting spies, said:

I met Yehuda early in my active career. Like many of my Colonel leagues, I was very impressed by him. He seemed to have a thorough understanding of the essence of operational intelligence . . . After I became aware of all the question marks surrounding the [Syrian] source, I decided to investigate further . . . The report of a surprise military attack that was supposed to take place in July 1996 turned out to be without any foundation. I witnessed how Yehuda Gil's false reports could have resulted in war . . . This made me realize that it was imperative to examine the reliability of the operative and not of the source . . . This incident illustrated the central role played by truth in the intelligence services' organizational culture. A number of operational activities finally brought us—from September to

November 1997—to realize that there was not and never had been a Syrian source. In November 1997 Yehuda was arrested . . . Proper organization must include monitoring, supervision, skepticism, and pessimism, the cornerstones of intelligence; in my opinion, if these exist, they can prevent such incidents from occurring. (Meltzer)

In the sphere of cooperation with allies, an example from Mossad activities can be presented. In the 1960s and 1970s, the Mossad maintained deep relations with the Kurds in Iraq, including military and civilian aid and political support. In this framework, the Mossad placed operatives in the headquarters of the Kurdish leader, Mullah Mustafa Barzani. They maintained close contact with the Kurdish leadership on a daily basis and invited them to participate in internal consultations. Eliezer "Gayzi" Tsafrir writes,

The two members of the delegation agreed to help in any way possible. At times, they made proposals on the spot. At others, they reached conclusions together [with the Kurds]. At other times, they sent queries to Israel for examination and analysis. Sometimes they said, 'Sorry; it is your decision. We aren't going to intervene; whatever you decide is okay with us.'

These three examples illustrate a culture of mission command, in which high-level professionalism is likely to lead—if employed properly—to striking achievements. They indicate the uniqueness of the Israeli case, in which professionalism is acquired through not only training but also the ability to improvise in the field and thus acquire valuable experience.

Nevertheless, the incidents described above belong to an earlier, pre-high-technology era. Elite forces now have technology enabling constant contact by satellite or internet, which is relatively secure due to encoding, making it possible to receive real-time streaming of what is happening on the ground, either from the force itself or from the means of the collection accompanying it, such as UAVs. For example, these capabilities played a prominent role in the U.S. Army's operation to eliminate Osama Bin Laden, as manifested in the pictures of U.S. decision makers watching and listening in the Presidential Emergency Operations Center. Such a reality means that elite operations, which can lead to either strategic achievements or entanglement, will be centralized in the hands of senior decision makers, both civilian and military. Technology emphasizes the paradox of mission command in intelligence operations that link the highest-quality operational units, whose judgment can be relied upon, and very sensitive operations that demand centralized control.

Mission Command and Intelligence

The "precision weapon revolution" demands that intelligence services supply real-time intelligence about the enemy in order to enable precise targeting, especially from the air. In 1987, an article published by Brig. General (Res.) Dr. Haim Ya'abetz, a senior ex-intelligence officer and advisor to RAFAEL Armament and Weapons Development Authority, summarized this process of change:

> The components of intelligence that serve military power, according to the [newly] recommended defense doctrine, must include intelligence closely connected with military systems based on advanced precision weapons launched from land, air, and sea.
>
> Passive and active intelligence systems enabling the integration of the air force in ground battles in an environment of massive enemy air defense and SSMs parallel to its mission to defend Israeli air space; and intelligence systems aimed at acquisition of targets, thus actualizing the full fire power potential in our possession.

Based on these capabilities, the I.D.F. in the 1990s adopted an approach promoting the accurate targeting of the enemy's military capabilities deep within its territory without crossing the border on the ground. This approach was based on the integration of precise weapons systems; advanced collection methods and targeting, command, and control systems; computerized communication; and intelligence systems (jointly known as C4I). These precise intelligence-based targeting systems have strengthened a centralized control of fire operations, in preference to maneuvering, which is essentially decentralized in nature.

Turning the intelligence service into an additional tool in the commander's hands has led to the establishment of the I.D.F. Intelligence Directorate Operational Division. In an interview with Brig. General K, head of the division and DJ2 for operations, he indicated that he was "responsible for the cruel prioritization of sparse [intelligence] resources in face of numerous threats." A triumvirate that includes the head of the Operational Division in the Operational Directorate (the DJ3) and head of the Air Force Air Group (DC IAF for Operations) does this. The target list is based on this prioritization, which enables prolonged fighting while collecting sensitive, precise intelligence data regarding leaders of terrorist organizations in order to target them as well. The Operational Division is responsible for operations and sorties in the intelligence sphere that are approved on a weekly basis by the commander in chief, the defense minister, and the I.D.F. target list. The control of a precise targeting intelligence "valve" has become a major component in the senior commander's toolbox, thus enabling centralized control of operations.

In recent years, the I.D.F. intelligence directorate has made additional strides toward applying this approach to ground maneuvers as well. Maj. General Aviv Kochavi, the former head of the directorate and the current Chief of Staff, describes the process of change taking place:

It has become apparent that the I.D.F. cannot realize its might against an enemy (that employs camouflage and hiding tactics and controls considerable fire power) except by specific, precise intelligence available to the forces on a micro-tactical level. This demands that the directorate must alter its activities and transfer the resources from information-Collection expeditions for other types of missions [to micro-tactical intelligence].

In order to realize an 'intelligence-based warfare' approach in cooperation with the ground forces, a number of activities have been planned, including a series of practical organizational changes: additional intelligence officers and desks were placed in the I.D.F.'s division headquarters; a greater intelligence presence was also planned (and later carried out) in the headquarters of maneuvering brigades . . .

Maj. General Kochavi understands that intelligence-based warfare, which contributes a wealth of accurate data enjoyed by firepower, also poses considerable challenges for the mission command approach. He goes on to describe this inherent paradox:

Intelligence-based warfare is not a trivial process, either for intelligence personnel or for the ground forces. It has raised and continues to raise questions such as the tension between the need for available intelligence and the friction and uncertainty that are always present on the battlefield . . . and the fear of the ground forces' excessive dependence on intelligence in battle conditions.

Intelligence-based warfare is situated between two extremes: on the one hand, as a means of turning maneuvering into a "factory" for targeting the enemy by maneuvering forces and close fire support, and on the other, as a tool in the tactical commander's hands, enhancing their creativity and use of stratagems. It appears that the "maneuvering targeting factory" approach has a greater chance of success than enhancing the commander's creativity, especially in light of the ground forces' growing jealousy of the attention given to aerial firepower (the "queen of the battle") and the political level's desire to avoid casualties at any cost. As stated before, the first option also has its cost in terms of mission command within the intelligence directorate, as it strengthens a tendency toward centralization as opposed to decentralization.

Conclusions

Mission Command and Intelligence:
How technology increases the Built-In Paradox

Figure 3: The technology paradox

The mission command paradox in intelligence can be found on all three levels: mission command in intelligence, mission command of intelligence, and mission command and intelligence. On the one hand, technology promotes mission command since it affords the average intelligence operative greater capabilities and networking possibilities within the intelligence directorate and with its clients. On the other hand, it provides the commander with tools that centralize their control over intelligence work.

The influence of technology on mission command—in everything relating to intelligence—can lead to two extremes. One extreme involves the use of quality intelligence to gain significant superiority in firepower and maneuvering for precisely targeting the enemy in such a way so as to protect creativity in coping with the uncertainty of strategic and tactical challenges. In such a situation, destruction of the enemy will be extensive enough to obliterate the element of surprise and "battle fog" at their disposal while also minimizing collateral damage and civilian casualties.

The other extreme is the use of precise intelligence in order to encourage new ways of using creativity and stratagems, a use characterized not by the flexibility of classical mission command but an advantage stemming from the speed measured in very short time intervals (minutes or even seconds), the superiority of information, and the availability of a variety of fighting capabilities.

Whatever the influence of technology and the evolving human factor on fighting, war will continue to be primarily a human phenomenon. The idea of mission command, which allows fighting to incorporate personal excellence for achieving military targets in uncertain conditions, will continue to influence warfare. Therefore, it is likely that armies will continue to find ways of implementing it in the future. The greatest challenge facing those shaping the future of warfare and the type of mission command that will be used is knowing how to best employ the intelligence directorate, which has become an indispensable supporting resource in the information revolution era.

Bibliography

Agranat National Commission of Inquiry. "A second interim report."
    January 7, 1975, p. 158 [Hebrew]
Barkai, A. "The Beating Wings of Error" Israel Intelligence, Heritage, &
    Commemoration Center (IICC). [Hebrew]
Barron, I. "Intelligence Research" Israel Intelligence, Heritage,

& Commemoration Center (IICC). [Hebrew]. Accessed
from: http://www.terrorism-info.org.il/Data/articles/
Art_20837/114_15_6816fifty 368.pdf.

Bar-Yossef, U. *The Observer Who Fell Asleep*. (Zmora-Bitan Publishing,
2001). [Hebrew]

Ben Porat, Y. *Closing Prayer*. (Idanim Publishing Tel-Aviv 1991).
[Hebrew]

Buchbut, A. The "godfather" of the Military Intelligence Directorate:
The shadow man who decides when to blow up Nasrallah's bunker.
Wallah. April 22, 2016. [Hebrew]

Brun, Itai. "Intelligence Analysis – Understanding Reality in an
Era of Dramatic Changes." Israel Intelligence, Heritage, &
Commemoration Center (IICC), The Intelligence Heritage Research
Institute, 2018, pp. 108–109. Accessed from: https://www.terrorism-
info.org.il/app/uploads/2018/08/114_15_E.pdf

Colonel, A. "Intelligence and Criticism." *Ma'arachotm*, 458. December
2014. [Hebrew]

Committee of Inquiry of the Yom Kippur War: Arguments and
completion of the partial report. April 1, 1974. [Hebrew]

Harel, A. "In 8200, they are praising the conscientious objector, but
criticizing his behavior." Haaretz, January 31, 2003. [Hebrew]

Kfir, A. & Dor, D. Barak: *The Wars of my Life*. Israel: (Kinneret Zmora-
Bitan Publishing. 2015). [Hebrew]

Kochavi, A. & Ortal, E. "The Military Intelligence Directorate's
Activities: Permanent Change in a Changing Reality." Between the
Polars, Second Issue. July 2014. [Hebrew]. Accessed from: http://
maarachot.idf.il/72255-he/Maarachot.aspx

Meltzer, E. (Ed.) *The Spy That Never Was and the War that Never Broke
Out*. Israel Intelligence Heritage & Commemoration Center (IICC).
January 2016. [Hebrew]

Siman Tov, D. & G. Ofer "Intelligence 2.0: A new Approach to

Performing Intelligence." *INSS Army and Strategy*, Volume 5. December 2014. [Hebrew]

Tsafrir, E. *Ana Kurdi*. (Ma'ariv - Hed Arzi Publishing. 1999). [Hebrew]

Ya'abetz, H. "Intelligence and the Security Approach to Force Build-up." In Zvi Ofer and Avi Kober (Eds.), *Intelligence and National Security*. (Ma'arachot Publishing, Second Edition.) [Hebrew]

Col. Shay Shabtai is an expert and practitioner for more than twenty-five years in Middle East issues, Israel's national security, intelligence and strategic planning. He served as the head of the Long Term Strategic Planning Department of the I.D.F.

Shay received his MA Magna Cum Laude from the Tel Aviv University Executive Program for Middle East studies, is a doctoral student researching the influence of Israel's intelligence community on the national security strategy, and lectures in Bar Ilan University and the IDC in Herzlia. His current research also includes alternative national strategies, strategic communication and perception management, cybersecurity strategy, and military transformation. He is the strategist of the Konfidas Cybersecurity Consulting firm, which works with Israeli banks, airline and shipping companies, and other leading firms.

Part 2

An Army is Born

1

Mission Command Follows the Army Built-Up

Brig. General Gideon Avidor

The present chapter surveys the development of battle doctrine as written doctrine; its stages of development; and how it is influenced by seasoned staff officers with operational experience. This was the period of an army under construction that came into being on the battlefield. Thus, the role of doctrine was to make official what was being done in the field. This is a departure from other armies, in which written doctrine precedes actual warfare.

Ben-Gurion, the prime minister and minister of defense at the time, wanted to model the army according to British and American models. Therefore, he recruited officers who had formerly served in those armies (a famous example being Col. David "Mickey" Marcus), mandating them to conduct a survey and recommend an organizational framework and battle doctrine. One of Marcus's major suggestions was to found a battalion commanders' school. Marcus was killed on 11 June 1948, and the school was not opened until 1949.

For political reasons, Ben-Gurion disbanded the various underground organizations and appointed officers who served in the Jewish Brigade during WWII. One of the key officers was Haim Laskov who was in charge of the commanders training program in 1949.

When doctrine is written based on performance, the commanders' personalities have an impact on its formulation. At that period, there were deliberations between an infantry-based army supported by other arms versus a deep, wide maneuvering approach based on armor. A strong supporter of the infantry approach was the then commander in chief, Moshe Dayan.

A striking example of mission command was the 1956 campaign in Sinai when, contrary to the G.H.Q. and Dayan directions, an armored brigade made a breakthrough, which dictated the campaign's outcomes. The I.D.F., trained for mission command, changed direction and armor— which, until then, was supposed to convey using tank transporters behind ground forces and only later break forward—became the chief arm on which the ground army was based.

Both doctrinal debates continued for another few years regarding the proper role of the tank as a major attack weapon or as a supporter of ground troops; the former approach won out and reached its height during the 1967 Six-Day War. In the aftermath of that war, doctrinal changes were made. Throughout these organizational pendulum swings, the dominant approach in matters of leadership and command stemmed from commanders' experience in the field. Thus, mission command remained a central tenet, at least in practice, while debate continued surrounding official written doctrine.

2

"They did it their way"
Mission Command in the I.D.F. 1949-1956
Dr. Dov Glazer

Over the past three decades, mission command has been attracting an increasing number of adherents, military practitioners, and academics alike. Its primary appeal lies in its promise to increase combat prowess, as its inherent flexibility enables a unit to adjust to the ever-changing battlefield. If the Prusso-German experience from Moltke the Elder onwards is any indication, a successful implementation of this command philosophy can provide significant dividends. After the colossal 1806 defeat at Jena and Auerstedt, the Prussian and later German armies demonstrated unparalleled tactical and operational prowess, at least partially due to implementing this command culture.

On a very basic level, mission command denotes a decentralized command system wherein a commander only designates an end goal to be achieved, allowing their subordinates the freedom to independently devise and execute plans to realize their objectives. It requires complete trust between commander and subordinate, born of shared study, experience and terminology. These encourage the subordinate's commitment to the commander's intent, and allows the commander to loosen the reigns. This command style depends on aggressiveness and initiative coupled, and complemented by, detailed systematic preparations. Furthermore, it requires first-rate staff work, from the initial evaluation of the situation through continued monitoring and coordination of efforts on the battlefield, as well as the design of future operations. The headquarters must be commander-centric, economical and flexible, well-defined and coordinated. When a commander prefers to command from the saddle on the front lines, staff members will also share the burden of making operational decisions. While relevant to all forms of war, this system is best suited for fluid, maneuver situations, as are inherent in mobile warfare and battles of encounter. As Eitan Shamir has shown, attempts to emulate this system have encountered significant challenges, primarily due to conflicting and deeply ingrained military cultures.[1]

This chapter seeks to examine more closely two of the aspects described above that are crucial to a successful adoption of mission command: the first is combat doctrine, and the second is the very nature of the command and control system, primarily that of the staff. It is not a theoretical treatise but rather a "nuts and bolts" examination of a concrete historical example: the Israeli experience from 1949 through 1956.

While the Israelis had not set out to adopt the Germanic approach to war, they nevertheless gravitated toward it, albeit unknowingly. Tracing their efforts to design a system of command and control to complement combat doctrine can illuminate the organizational, cultural, and theoretical forces at play in a military establishment stumbling—and at times fumbling—its way toward its own unique version of mission command.

During that period, doctrinal development in the I.D.F. was carried out by the Instruction Department through four of its leading agencies: the activation team, the battalion commanders' course, the Command and Staff College, and the infantry department. The present analysis will follow these agencies, while offering a critical examination of the doctrine they produced in relation to four pivotal aspects: command, operations, combined arms, and a basic understanding of doctrine. The primary question is to what extent could the Israeli command and staff structure and combat doctrine between 1949 and 1956 support mission command? As the following pages will demonstrate, the Israelis adopted a maneuver-oriented doctrine early on, drawing heavily from Anglo-U.S. doctrine

1   Eitan Shamir, *Transforming Command: The Pursuit of Mission Command in the US. British and Israeli Armies* (California: Stanford University Press, 2011).

but resembling the Germanic experience. Also, the men who adopted, adapted or developed this doctrine where mostly combat commanders. However, the model Israel developed proved more restrictive in various ways, not the least of which was moving away from combined arms and a structure-of-command that hampered its effectiveness and ability to realize mission command. To paraphrase Sinatra: they did it their way.

The Challenge

During the three decades of the British Mandate in Palestine, the Jewish community supported three primary underground militia organizations: the Haganah (1920–1948), Irgun (1931–1948), and Lehi (1941–1948). These, along with thousands of British, Soviet, and U.S. World War II army veterans, conjoined to give rise to the I.D.F. on the battlefields of the 1948 war. The Jews won that war, defeating the military organizations and irregular forces of the Arab community in Palestine (the Arab Liberation Army and the Egyptian Army) while also keeping the Jordanians, Syrians, and Iraqis at bay. By the summer of 1948, the I.D.F. possessed a battalion-level combat doctrine, the precursor of a true Israeli combat doctrine. Half a year later, through superiority in numbers and materiel—as well as command and command functions—the Israelis achieved strategic decisions on the battlefield. By January 1949, the I.D.F. had passed the ultimate test, having ensured the coming into existence of the new state of Israel. With the onset of ceasefire negotiations, the I.D.F. faced a new challenge: that of maintaining its hard-won edge in peacetime. Demobilization—leading to an inevitable downgrading of operational capabilities and coupled with new strategic constraints—required a new approach.

Following, to some extent, the Swiss model, the army's new battle order included a regular component made up of two-to-three year conscripts, a small permanent component, and a large reserve force[2] organized in

2   Yitzhak Greenberg, *The Israel Reserve Army* (Bear-Sheba: Ben-Gurion University Press, 2001). [Hebrew]

reinforced brigades according to the British model. The post-war army would peak at 40 percent its wartime size, shrinking from 100,000 to a mere 40,000 officers and men.[3] Moreover, many of the best and the brightest were retired, forcibly or voluntarily. Few of the new recruits, mainly recent immigrants from North African countries, possessed the language skills, technical and practical understanding, or motivation that had characterized their predecessors.[4] For instance, in an open discussion at the end of a battalion commanders' course, the participants indicated that a combat commander's primary concern at the time involved welfare issues.[5]

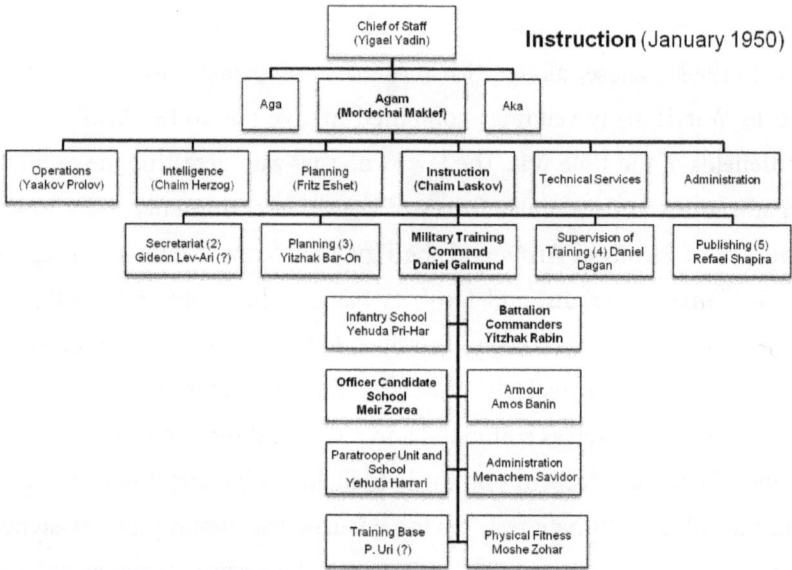

Table 3: Instruction Department (January 1950)

By the spring of 1949 with the changing of the guard, the I.D.F. General Staff was in the throes of a major reorganization. Styled after the

3   Ze'ev Elron, "The Development of the I.D.F. and the Change that never was in Israel's Strategic Doctrine, December 1952–September 1955." (PhD Dissertation, Hebrew University, 2009), p. 17. [Hebrew] Needless to say, this change met with considerable opposition within the I.D.F. op. cit., pp. 27–34.

4   Sagi Torgan, "Training I.D.F. Combat Commanders, 1949–1956" (PhD Dissertation, Hebrew University, 2008), p. 65. [Hebrew]

5   Summation of the Course for Battalion Commanders, 4 April, 1949, pp. 9–17, I.D.F. and Defense Establishment Archives (hereafter cited as I.D.F.A) 96-854/1952. [Hebrew]

British General Staff, , Israeli G.H.Q (General Headquarters) was reduced to three (G, Q, P) directorates: General Staff (J3), Quartermaster (J4), and Personnel (J1).[6] In the new organizational structure, Instruction was reduced from an independent Directorate to a department, subordinate to Staff (Operations J3).

This last precept, according to which Israeli commanders are mentioned by name, has paved the way for another one. Ever since its inception, and at least until the late 1950s, Instruction was staffed primarily by combat commanders and was commanded by future senior commanders. Doctrine, at least during that timeframe, reflected field-grade thinking. Moreover, most of those individuals had actually passed through all four Instruction agencies mentioned above. Its functions were distributed among five, and for a brief time eight, branches and two services, as well as the paratrooper unit.[7] Its office for military schools, the Military Training Command, established several key schools in the autumn.[8] This organizational chart remained in effect until December 1953, when incoming Chief of Staff Moshe Dayan appointed Yitzhak Rabin, a 1948 war brigade commander and recent graduate of the British Command and Staff College at Kimberley (December 1952–December 1953), to the department and re-elevated it to a General Staff directorate.[9]

6   While discussing the structure of the general staff in 1951, the Israelis translated excerpts of Marine Lieut. Col. J. D. Hittle's, *The Military Staff: Its History and Development*. I.D.F.A 186-433/1956. [Hebrew]

7   Doctrine department also briefly included two services: a physical fitness service (soon transferred to the Infantry School) and a canine service (transferred to the police). For a detailed organizational chart of Doctrine Department: Office of the Head of Lahad: Structure of Pahad, I.D.F.A 87-315/1953. [Hebrew] At one point, the department was divided into eight branches. Matcal/Ahad, establishment and liquidation orders, I.D.F.A 171-854/1952. [Hebrew] The names and purview of the branches were changed as well. Office of the Head of Ahad: Establishment and Liquidation of Units, January–April 1949, I.D.F.A 171-854/1952; Office of the Head of the Ahad: Structure of Military Training Command, I.D.F.A 87-315/1953. [Hebrew]

8   Battalion commanders' course (February), administration (31 March), infantry officers (31 March), infantry (12 April). Matcal/Operations, "I.D.F. Structure and Organization," I.D.F.A 22-854/1952. [Hebrew]

9   Several minor changes occured. For instance, a Doctrine department 5's section (from August 1950) became Doctrine department 6 (on 1 June, 1951) to handle the growing number of officers pursuing higher military education. Office of the Chief of Staff: Reports of the Committee for Continued Education Programs, In-country and Abroad, I.D.F.A 129-854/1952. [Hebrew]; Office

The organizational transformations detailed above were significant, as, since its inception in 1939, Instruction never boasted more than two dozen members. Over an 18-month period, it had grown in size and complexity, necessitating a complete overhaul of the way it conducted its business. This form of institutionalization, a prerequisite for training dozens of thousands of men and women to perform myriad military professional specialties, demanded knowledge and experience that was unavailable to the Haganah. It was established by ex-British army officers, who dominated the Instruction branch during the war and immediately afterwards. Before handing over the reins to 1948 war veterans, they performed one final task: standardizing the doctrine, training, and structure of the I.D.F.'s reinforced brigades.

The Haganah was essentially a collection of independent territorial commands. Efforts to standardize and coordinate training, weapons procurement, and planning met with varying degrees of success. The authority of the I.D.F.'s G.H.Q. strengthened during the 1948 war, but the brigades retained their pre-war independence vis-a-vis doctrine—as was the norm, for example, in the Kaiserheer corps system—force build-up and training. The tumults of post-war reorganization offered a unique opportunity to address these issues, and Haim Laskov of the Doctrine Department established a special activation team for that purpose. A combat commander, Laskov was also an educator and talented administrator. He saw action as a company commander in Italy during WWII and as battalion commander in the war of 1948.

Following his time in Instruction, Laskov would go on to command the Air Force (1951–1953), study at Oxford (1953–1955), become Head of Operations and Deputy I.D.F. Chief of Staff (1955–July 1956), commander of the Armored Corps (July–November 1956), a formation (Division)

of the Chief of Staff: Appointments and Vacations, April-November 1951, I.D.F.A 148-1559/1952. [Hebrew] In January 1953 the secretariat was liquidated and its responsibilities redistributed among the other branches. Office of the Head of Ahad: Summary of Branch Chiefs' Meetings, I.D.F.A 29-433/1956. [Hebrew] Also, in December 1950 Doctrine department 1 became the research branch. I.D.F.A 87-315/1953. [Hebrew] commander in the Sinai War (October 1956), commander of Southern Command (1956–1957), and finally, I.D.F. Chief of Staff (1958–1961).

Established on 29 August 1949 under Instruction command,[10] the activation team conducted an extensive survey—from 1 September 1949 to 10 February 1950—of the brigades' combat readiness. Deficiencies were found in all areas, from ammunition storage to worker allocations to properly defining roles and positions.[11] The primary deficiency was due to insufficient and inconsistent training. To address this issue, the team produced thirty-three training manuals,[12] mainly for the infantry, the backbone of the Israeli Army. One of these publications was *Training the Reinforced Infantry Brigade* (September 1949). The following discussion analyzes that field manual along with the doctrinally-similar *Brigade Operations' Staff Officers* (October 1949) and *The Reinforced Infantry Brigade* (March 1951).[13]

Training the Reinforced Infantry Brigade was the first Israeli manual devoted to brigades, although the Israelis had been deploying such formations for two years. Devoted to tactics rather than training procedures,[14] it foresaw a reinforced infantry brigade operating as a divisional maneuver element—independently or divided into ad hoc combat teams—performing a range of missions, from holding territory

10   Matcal/Operations-Matam, I.D.F. Structure and Organization, I.D.F.A 22-854/1952. [Hebrew]

11   "Termination of the Activation Team's Work, 13 February 1950," I.D.F.A 15-1166/1951. [Hebrew] Regarding the difficulties, see for example: Office of the Head of Doctrine department: Pahad Progress Reports, March 1948-November 1949, I.D.F.A 152-854/1952. [Hebrew] and: Chaim Laskov, "Final Report of the Activation Team, February 1950", pp. 50-51, I.D.F.A 41-61/1952. [Hebrew]

12   Laskov, "Final Report," pp. 51-52, 67.

13   Operations/Doctrine department, Brigade Operations' Staff Officers (October, 1949). [Hebrew]; idem, Infantry Tactics 5: The Reinforced Infantry Brigade (March, 1951). [Hebrew] The doctrinal similarity to the other manuals is evident, for instance. in the treatment of attack and defense: Operations/Doctrine department, Infantry Tactics 3: The Infantry Company (November, 1950), pp. 100-35, 172-88. [Hebrew]; Operations/Doctrine department, Infantry Tactics 4: The Infantry Battalion (January, 1951), pp. 74-114, 152-175. [Hebrew] Interestingly, The Infantry Company devotes an entire chapter to the platoon. The Infantry Company, pp. 60-91.

14   Operations/Doctrine department, Training the Reinforced Infantry Brigade (September, 1949). [Hebrew] as opposed to Operations/Doctrine department, Training the Infantry Battalion (August, 1949). [Hebrew], which is actually a training manual. The term was dropped in the 1951 edition.

to containing and destroying enemy forces to missions deep behind enemy lines.[15] Interestingly, references to the division echoed an ongoing debate within the General Staff about whether the I.D.F. should adopt such formations.[16]

Reflecting a "maneuvers" approach that emphasized initiative and speed over firepower, the manual subordinated the fire plan to the needs of the maneuvering element, that is, the infantry, and advocated a "fire and maneuver"-based attack. This was because fire facilitates maneuvering by forcing the enemy to take cover and inflicting casualties, thus allowing the maneuvering element to advance sufficiently for shock action. Interestingly, this manual was the only one of its kind to recognize the meeting engagement (to which the Germans had been devoting considerable attention starting in 1888)[17] as a type of attack and to advocate initiative and aggressiveness over caution and firepower in such cases.[18] It also adhered to the "maneuvers" style by defining brigade objectives in geographical terms, generally in terms of crucial areas deep behind enemy lines, the capture of which would result in the enemy's moral and physical collapse.[19]

As a rule, tanks were attached to infantry units and employed against enemy infantry, but they could also be deployed independently.[20]

The brigade attack is comprised of "primary" and "secondary" efforts,

15   Training the Reinforced Infantry Brigade, pp. 1, 35, 53, 57. As the I.D.F. only adopted division-like commands (Ugdot) in 1954, this stipulation is probably indicative of foreign origins.

16   Concerning this debate see: Ze'ev Elron, "Infantry in the I.D.F., 1949-1956" (MA Thesis, Hebrew University, 2001), pp. 36-37. [Hebrew]

17   Exerzir-Reglement für die Infanterie (Berlin: E. S Mittler und Sohn, 1888).

18   The Infantry Battalion, pp. 111-15, discussed the meeting engagement more rigidly, and at the end of the chapter, perhaps due to differences in force size, spatial dispersion and reaction-time cycles between brigade and battalion.

19   Training the Reinforced Infantry Brigade, pp. 35-38. The fire-support element also included close air support, the principles of which seem to have been drawn from: Operations/Doctrine department, "Air Force and Ground Force Cooperation," Training Bulletin 10 (July, 1949), I.D.F.A 204-433/1956. [Hebrew]

20   The Infantry Battalion, pp. 102-05, allotted a more extensive role to armor in a discussion resembling: Operations/Doctrine department, Tactical Employment of Infantry-Armor Operating in Cooperation (May, 1951), I.D.F.A 27-1629/1952. [Hebrew]

and relies on stratagem and envelopments.[21] As attacks rarely developed as planned, the manual encouraged commanders to adhere to the plan aggressively, but not blindly; overcome unforeseen circumstances; exploit unanticipated success; and reinforce failure only in order to minimize casualties or check enemy movement.[22]

Timely reports enable commanders to continuously evaluate possible enemy courses of action and thereby prepare appropriate countermeasures. Accordingly, the 1951 edition added a list of five governing principles: maintenance of mission (defined as destruction of the enemy); surprise; mobility; concentration (optimizing available forces); and combined arms and ground-air cooperation. It also emphasized the significance of reconnaissance and stratagem.[23]

The envisioned defense was aggressive, dynamic, and mobile, resembling the WWI German doctrine known as the *Abwehr im Stellungskrieg*.[24] This doctrine relied on consecutive zones established in depth, beginning with observation posts in the outer zone, followed by the main force and being concluded by the reserves. Defensive arrays, as the manual stated, revolved around a crucial point, the fall of which would result in the collapse of the entire array and necessitate a fierce defense.[25] This was the closest any contemporary Israeli (or American) manual had come to recreating the original German doctrine. Subsequent manuals neglected some of its defining characteristics, especially counterattacks.

21 Training the Reinforced Infantry Brigade, pp. 38-39; The Reinforced Infantry Brigade, pp. 116-20. Ibid., pp. 123-25 includes an identical paragraph on "primary" and "secondary" efforts, but only later on in the chapter. The decision to insert it in the opening paragraph reflects the importance attributed to it. In contrast, The Infantry Battalion, pp. 68-69, 89-90, discussed only the primary effort, signifying the declining significance accorded to stratagem.

22 Significantly, this truism only appears on page 45 rather than in the very first paragraph, as was common, for example, in German manuals.

23 Training the Reinforced Infantry Brigade, pp. 45-46; The Reinforced Infantry Brigade, pp. 112-14.

24 Several authoritative accounts have been published on the topic, including Timothy T. Lupfer, "The Dynamics of Doctrine: The Change in German Tactical Doctrine During the First World War," Leavenworth Papers 4 (Fort Leavenworth: Command and General Staff College, 1981); Martin Samuels, Command or Control: Command, Training and Tactics in the British and German Armies 1888-1918 (London: Frank Cass, 1995).

25 Training the Reinforced Infantry Brigade, pp. 58-59.

While recognizing the need to dedicate combined-arms units for counterattacks utilizing surprise, daring, speed, and timing, training the reinforced Infantry brigade alone recognized that counter-attacks were demanded on all levels of command; the counterattack force constituted fully one third of the unit. Having said that, even this manual fell short of the Prussian-German model by resting authority for launching the counterattack with the brigade commander rather than with the counterattack detachment's commander, who was obliged to submit his plan for approval.[26]

Leadership is exercised by example. Commanders inspire and encourage decisiveness, aggressiveness and innovation.[27] The commander must be familiar with the technical and tactical aspects of all the forces and weapon systems at their disposal. They devise the operational plan and remain responsible for its execution, while granting subordinates full independence, intervening only to ensure cooperation between units or to avoid glaring mistakes.[28]

The command post should facilitate control over the brigade,[29] and the H.Q. should enable control and communications, while affording staff sufficient accommodations as well as concealment.

To minimize disruption of staff work in a crisis, alternative locations should be scouted out in advance, and staff should be organized in shifts to enable 24-hour operations.[30] Brigade H.Q. is divided into forward

---

26   Op. cit., pp. 60, 70, 72. In addition, the principles of the counter-attack are discussed in the opening paragraph of the chapter on defense, rather than being relegated to Paragraph 8 as in: The Reinforced Infantry Brigade, pp. 177-82. The revised attitude toward counter-attacks was already anticipated in: The Infantry Company, pp. 167-68, 181-83; The Infantry Battalion, pp. 172-76.

27   Training the Reinforced Infantry Brigade, p. 1. Compare: The Infantry Battalion, pp. 8-9, 48. These principles apply to all levels of command, as demonstrated by the range of topics covered in: Operations/Doctrine department, Infantry Tactics 1: The Infantry Squad (August, 1951). [Hebrew], including: intelligence, armor, communications, attack and defense.

28   Other manuals require commanders to be ready to assume command of the next higher formation. The Infantry Company, pp. 55-56; The Infantry Battalion, p. 69.

29   Training the Reinforced Infantry Brigade, pp. 2-3, 22, 45, 49.

30   Training the Reinforced Infantry Brigade, pp. 20-1. However, the manual fails to specify how this can be achieved. For instance: who should replace the operations officer - especially as the individual best suited for the job, his deputy, has his own permanent responsibilities?

(tactical) and rear (logistic) elements. However, since the size of the tactical element—including the commander, operations officer and their deputy, intelligence, administration, technical, military police, medical, headquarters commander, signals and fire-support—impedes mobility, the manual advanced the idea of a command group comprising only the commander and a few aides.[31] Staff work is coordinated and supervised by an operations officer who is privy to the commander's intentions.[32]

Orders, the basis for cooperative execution of the operational plan, should be concise, simple, and carefully worded, relying on shared terminology and graphic content when appropriate.

Orders include an analysis of the terrain; the intelligence reports upon which the operational plan is based in order to facilitate adaptation to unexpected circumstance; disposition of friendly forces; and intent, through which the commander divulges their plans for the next stage. The method provides answers to the questions, "who, what, when, how and why." Orders also include the fire plan that describes the primary effort and the deception plan, as well as logistics and administration.[33]

The staff structure and responsibilities described above were revised a month later (October 1949), in the *Activation Team's Brigade Operations' Staff Officers*. It introduced a new position to the organizational chart, namely, the deputy commander, conferring upon them many of the operations officer's duties.[34]

31   Training the Reinforced Infantry Brigade, p. 10; The Reinforced Infantry Brigade, pp. 13, 35, 49-53. The Infantry Battalion, pp. 12-13, adopted the same configuration, even though the battalion staff is considerably smaller, making this division cumbersome. Interestingly, the manual begins by raising objections to the idea, then devoting the majority of the discussion to the proper application of this division.

32   Training the Reinforced Infantry Brigade, p. 4-5.

33   Op. cit, pp. 15-6. Absent from this manual, but rectified in the 1951 edition, is a discussion of logistics. The Reinforced Infantry Brigade, pp. 54-80. The new manual established that due to its bearing on the operational plan, the logistic plan also falls under the commander's purview. To support mobility and maneuvering, the plan must be simple and flexible, aiming to meet rather than exceed operational needs, in order to avoid straining their transport capabilities. For the same purpose, supplies should be pushed forward rather than pulled by the recipients. Op. cit., 54-6.

34   Brigade Operations' Staff Officers, pp. 1, 12, 20. According to: Hanan Shai (Schwartz), "Generalship in Israel's Wars and the Art of War" (PhD Diss., Hebrew University, 1996), p. 131. [Hebrew], the position of a deputy, emulating the American staff system, was introduced under

In addition, it revised the purview of various staff officers, primarily that of the operations officer and their deputy and it standardized the three-headquarters division (rear, main, and command group). More notably, the manual transferred the responsibility of formulating the order from the commander to their operations officer,[35] an innovation that was adopted by all subsequent manuals.[36] While seemingly innocuous, the alteration reflected the confusion—which persists to this very day—regarding the purview and duties of the commander, deputy commander, and the operations officer.[37]

Firstly, both deputy commander and the operations officer are expected to serve as de facto chiefs of staff, on top of other duties. For instance, the deputy is authorized to command a brigade element or supervise the rear headquarters. Clearly, staff work cannot be properly coordinated under such conditions.[38]

Secondly, since the orders reveal the operational design—and the commander's thoughts and will subsequently guide the subordinates' adaptation of the plan to developments on the battlefield—it is the commander who must write the order, or at least the intent paragraph, that defines the operation's desired result. Acknowledging this point, the manual later added:

> Intent must be clear and take the form of a definitive and imperative sentence, explicating the mission the commander's intent and

---

advice of the American volunteer, Col. (Ret.) Fred (Gronich) Harris. About Harris see: Tom Segev, 1949: The First Israelis (Jerusalem: Keter, 2001), pp. 251-56. [Hebrew]

35   Brigade Operations' Staff Officers, pp. 2, 16-7, 19, 21-3.

36   The Reinforced Infantry Brigade, pp. 16-8. The entire section devoted to the staff seems to be a reproduction of the September/October 1949 manuals. In issues such as the duties of the deputy operations officer, the 1951 edition follows the October manual more closely. Compare for instance: The Reinforced Infantry Brigade, 19; Training the Reinforced Infantry Brigade, pp. 6, 8; Brigade Operations' Staff Officers, p. 23.

37   For example, Boaz Zalmanovitch, "Who needs a Chief of Staff?", Ma'arachot 435, February 2011, pp. 68-71. [Hebrew]

38   The possibility of their working in shifts, allowing the staff to maintain a 24-hour cycle is negated by the assignment of permanent duties. Essentially, they both have to be awake at the same time.

clarifying to the recipient what he must do. The commander must phrase and dictate it himself. The mission will be based on . . . intelligence reports. Next, it is important to indicate the mission that will follow, unless it impedes understanding of the primary mission or obscures it.[39]

These conflicting views were not reconciled. Interestingly, "corrections" appended at the end of the manual replace the term "intent" with "mission," changing the wording of the paragraph accordingly. This change effectively shifted the focus from the intended result to the mission itself, later requiring subordinate commanders to adapt their plans according to the initial mission, rather than to the intent behind it.[40]

From the perspective of mission command, Israeli Doctrine travelled an uneven path.

The combat doctrine was an essentially "maneuvers" and decentralized approach, as demonstrated by a policy of non-interference on the part of commanders; the continual emphasis on aggressiveness, initiative, and deception; reliance on combined-arms formations and combat teams; subordination of the fire plan and the logistic apparatus to the needs of the maneuvering element; the defensive array's dynamic nature; the designation of objectives deep behind enemy lines;[41] and so on.

However, certain factors impeded this approach from attaining fruition. The first of these was the size of the tactical H.Q. that restricted its mobility, thereby forcing the commander to choose between two extremes: either relinquishing control and allowing their subordinates free reign (umpiring) or tightening their grip (centralized command). The second factor was adopting the U.S.-based position of deputy commander, while according the operations officer similar responsibilities, as in the

39   The Reinforced Infantry Brigade, p. 43.

40   Op. cit, p. 238. Compare: The Infantry Company, pp. 71-2, 116-18; The Infantry Battalion, pp. 40-5.

41   For a bolder and more daring approach: Operations/Doctrine department, Deep Infiltration (June, 1952), p. 3. [Hebrew] See also: Ze'ev Elron, "I.D.F. Infantry: 1949-1956" (M.A. Thesis, Hebrew University, 2001), pp. 26-7. [Hebrew]

British-German case. By authorizing both of them to coordinate staff, the new system seriously injured staff work. Thirdly, the gradual transfer of responsibility for writing orders (especially the "intent" paragraph) from commander to staff reflected a shift from emphasizing the objective to the mission itself.

Moreover, it obfuscated the chain of command, also shifting the focus of command and control from the commander to the staff. Considerable attention was paid to infantry-fire-support cooperation and all-arms combat teams, as well as to a fluid and cooperative, rather than a rigid deployment of these forces. However, the armor essentially was relegated to support the infantry, reflecting a restricted view of armor in particular and combined-arms formations in general. The doctrinal approach was descriptive; that is to say, it relied on guidelines or suggestions rather than on dictating courses of action. In addition, in many instances, the discussion centered on the advantages and disadvantages of various alternatives.[42]

The issuing of these manuals also signified an end of an era, a changing of the guard. In August 1951, Haim Laskov, the driving force behind the Doctrine Department, was appointed commander of the air force.[43] Meir "Zarro" Zorea replaced him in the Doctrine Department. Having served as a company commander in the British army in Italy, Zarro was a battalion commander (infantry) in the First Arab-Israeli War (March–December, 1948), commander of the officer candidates' school (January–December 1949), and a student at Kimberley (1951) prior to this assignment. Following his tenure at the Doctrine Department (September 1951–December 1953) and living a few years as a civilian, Zarro served as assistant to the Head of Operations (February–

42   As with the discussion of the night attack: Training the Reinforced Infantry Brigade, pp. 50-2, or the counter-attack: Op. cit, p. 70. It is noteworthy, that while night attacks are valued, they are not yet considered a preferred method of attack. Although according to Elron, "I.D.F. Infantry", pp. 23-4, it was the I.D.F.'s modus operandi.

43   Office of the Chief of Staff: Appointments and Vacations, April-November 1951, I.D.F.A 148-1554/1952. [Hebrew]

November, 1956), Head of Operations (September 1958–April 1959), Head of the Doctrine Department (March–September 1958), and, finally, as commander of the Northern Command (April 1959–July 1962). Zarro committed the Doctrine Department to two primary tasks: implementing the activation team's recommendations and establishing the I.D.F.'s Command and Staff College.

Doctrinal development during his tenure centered on such issues as ground-air cooperation, the influence of climate and ground factors, tactical symbols, and so forth.[44] If the articles penned by his officers in the army journal *Ma'arachot [Campaigns]* are any indication, the Doctrine Department invested considerable resources in devising proper training methods.[45] As a testament to his beliefs, when Erwin Rommel's WWI memoirs *Infanterie greift an* were published in Hebrew (1953), Zarro ordered every officer to read it, to learn "leadership, caring for men, aggressiveness in every situation, initiative, and resourcefulness."[46] During that period, Doctrine Department branch chiefs included a future chief of research in the Intelligence Directorate (David Carmon), two Six-Day War brigade commanders, (Eliezer Echt (Amitai) and Moshe Lantzet (Yotvat)), while school commanders included such future leaders as Israel Tal, Issachar "Yiska" Shadmi, Haim Bar-Lev, and Yehuda Wallach.[47] Wallach later became a renowned Israeli military historian.

That generation of Doctrine Department officers, veteran company, and battalion commanders of the First Arab-Israeli War were nearly all graduates of the Haganah's platoon commanders course. More importantly,

44   Hanoch Patishi, Meir Zorea: A Biography (Tel Aviv: HaKibbutz HaMeuchad, 2013), pp. 163-4. [Hebrew]

45   See for instance articles by Zaro, Gideon Rothschild, Alex Eliraz and Moshe Efron in 1953 Ma'arachot issues.

46   "Training instruction No. 16/11/53," I.D.F.A 20-636/1956. [Hebrew] By and large, this description is validated by Doctrine department weekly/monthly meetings. Office of the Head of Doctrine department, Summary of Branch Chiefs' Meetings 1953-1954, I.D.F.A 29-433/1956. [Hebrew]

47   He was replaced in the 5th Brigade by Zvi Zamir, whose place Wallach occupied in the course. Office of the Chief of Staff: Appointments and Vacations, April-November 1951, I.D.F.A 148-1554/1952. [Hebrew]

nearly all of them had passed through the I.D.F.'s advanced officer training and education center, the post-war hothouse for doctrinal development.

Higher Education in the I.D.F.

In mid-February 1949, as the shooting ended, the I.D.F. was prepared to establish the long-awaited course for higher commanders, the battalion commanders' course, which was the brainchild of U.S. Col. (Ret.) Mickey Marcus. A Jewish-American volunteer, Marcus had, in the spring of 1948, recommended that the I.D.F. establish such a course, offering to command the course and write instructional material for it together with Haim Laskov. However, commanders could not be spared in wartime for the duration of the course, so the plan was shelved for a year.

Nominally commanded by Laskov, actual course direction fell to Yohanan Peletz, Moshe Dayan's deputy battalion commander of the 89th Battalion (the Raiders). The first class (February–April 1949) boasted six instructors (four of whom were British army veterans) and twenty-seven students, including the future commander of the Southern Command, Asaf Simchoni.

The discussions held at the end of the course revealed several significant deficiencies: prospective students were lacking in professional knowledge and experience; insufficient resources were available for demonstrations (due to the need to maintain combat readiness); physical fitness was not optimal; transportation was lacking; supplementary materials an various languages were necessary; and more.[48] The second class began four months later, on 8 August 1949, after intensive preparations, to standardize the instructors' professional level.

Its chief instructor was Yiska Shadmi, who had taught the previous course. Other instructors included Aharon Yariv (a graduate of the previous course), Yeshayahu Gavish, Itiel Amichai, Haim Bar-Lev, Zvi Zamir, Uzi Narkiss, Israel Tal ("Talik"), Uri Ben-Ari, and Moshe Kalman.

48   Laskov's remarks in the concluding discussion at the end of course (4 April 1949), I.D.F.A 96-854/1952. [Hebrew]

Eight of the ten instructors were ex-Palmach members (Haganah assault companies) and the remaining two were ex-British army veterans. The student body included a future wartime chief of staff (David "Dado" Elazar), division commander (Yehuda L. Wallach), and brigade commander (Shmuel Galinka).

The material, based on British army manuals, included the following: battalion and brigade structure and organization; basic field craft (maps, topography, communications); preparation of fire-support plans; cooperative (anti-aircraft, sappers, armor) and joint (aerial and naval forces) operations; battle forms (attack, meeting engagement [battle of advance]); defense, pursuit, counter-attack and retreat; battle techniques (advance guard, urban warfare, guerrilla); principles of war; mission orders; situation estimate; battle procedures; standing orders and lessons of the war.[49]

The curriculum, as far as professional military issues were concerned, remained essentially unaltered over the following years, despite frequent changes in command (Yohanan Peletz was replaced by Yitzhak Rabin, then Haim Bar-Lev, then Yehuda Wallach, and finally Zvi Zamir). The materials were drawn from British army manuals, though the course also incorporated lessons learned in the 1948 war. For instance, when Rabin discussed the principles of war, his examples were drawn almost entirely from the 1948 war, while Chief of Operations Yigael Yadin focused on deficiencies in the I.D.F. performance during the war.[50] The course was replaced by the newly established Command and Staff College in 1954.

Initially, the Israelis could only offer sporadic seminars for senior commanders.[51] The success of the second and, especially the third,

49   Torgan, "Training," p. 201.

50   Amiad Brezner, Wild Broncos: The Development and the Changes of the I.D.F. Armor 1949-1956 (Tel Aviv: Ma'arachot, 1999), pp. 100-1. [Hebrew] In some matters, the school followed official I.D.F. doctrine. For instance: Battle Procedures in the Infantry Battalion, 1952, pp. 14-5, I.D.F.A 407-147/1961. [Hebrew]

51   "Organization of the Inter-Service Task Force Headquarters, [date unknown]," I.D.F.A 6-1291/1951. [Hebrew] see also: Office of the Head of Ahad, School HQ Progress Report, March 1948-November 1949, I.D.F.A, 152-854/1952. [Hebrew]

battalion commanders' course led to a decision to institutionalize the professional knowledge of brigade commanders, most of whom had only graduated from the Haganah's platoon commanders' course. The first "Phase B advanced officers' course" opened on 26 December 1950; its twenty-three senior officers spent the first two months undergoing an abbreviated "Phase A advanced officers' course."

The instructors were Zvi Zamir, Itiel Amichai (who left in April to join the Doctrine Department), Dado, and Uri Ben-Ari. In June 1951, Dado became chief instructor until he was replaced in June 1952 by Uzi Narkiss. Among the twenty-three senior commanders were front commanders Moshe Dayan, Yosef Avidar, Zvi Ayalon, and Moshe Zadok.

By the time the second course began in the summer of 1952, new instructors had joined the ranks: Wallach, Avraham Yoffe, and Ephraim Shorer. They soon were joined by Mattityahu Peled and Yehuda Prihar.[52]

As this long list demonstrates, the elite of Israeli mid-level and senior commanders passed through the halls of the school for battalion commanders. Most of them soon found their way to instruction.

Some of the doctrines developed in the course concerned armor, which took a circuitous route.[53] In late February 1948, Yitzhak Sadeh established the Armored Service, still as an underground Haganah unit. However, it was soon disbanded, as its men were reassigned to establish the new 8th Brigade (motorized) in the face of the Arab regular armies' invasion. At about the same time, the men of the Instruction Directorate formed an armored car battalion commanded by Haim Laskov as part of

52   Torgan, "Training", pp. 212-3. Pri-Har's former posting was commander of 10th Reserve Infantry Brigade. In July 1956 he became commander of the Command and Staff College. Head of Operations, Appointments and Cancellations, January-June 1956, I.D.F.A 47-1034/1965. Peled finished a term as chief instruction in Pum before joining it.

53   Yoav Gelber, The Emergence of a Jewish Army: The Veterans of the British Army in the I.D.F. (Jerusalem: Yad Ben Zvi, 1986), pp. 520-4. [Hebrew]; Moti Golani, There will be War next Summer...: Israel on the Path to the Sinai Campaign, 1955-1956 (Tel Aviv: Ma'arachot, 1997(, pp. 216-23. [Hebrew]; Mordechai Naor, Laskov (Tel Aviv: Ministry of Defense, 1988), pp. 253-58 [Hebrew]; Edward Luttwak and Daniel Horowitz, The Israeli Army (New York: Harper & Row, 1975), pp. 126-32.

the newly formed 7th Brigade (motorized) (28 May).[54] In the aftermath of the war, the 8th Brigade became a reserve infantry brigade, and the 7th Brigade became a regular reinforced armored brigade. When the activation team's armored committee developed doctrine for the Independent Armored Brigade, its principles were as follows:

2. The Armored Corps attacks in columns of combat teams by deep insertions over one or multiple axes of advance. 3. The Armored Corps will endeavor to destroy the enemy's flanks and depth; the armor strikes at "soft" areas [. . .] 7. The Armored Corps will not divide into small units intended to stop or delay the breakthrough of enemy armored forces. The armored corps will avoid armored confrontations.[55]

The doctrine presented two themes that would characterize I.D.F. armored doctrine in the coming years: deep thrusts behind enemy lines (as practiced, for instance, in the annual maneuvers held in 1951, 1952, and 1954) and a desire to avoid tank battles. Armor was regarded as a "soft" offensive tool, unequal to heavy tasks, such as breakthrough operations. The original desire to concentrate armor was abandoned in the early 1950s in favor of small units assisting infantry formations.[56] A certain improvement in the armor's status and development came as a result of the formation of a new professional position, namely, a chief armor officer (February 1950) who was incorporated within Instruction. The first year and a half proved tumultuous, as frequent replacements at the helm negated any long-term processes, such as doctrinal development, and the new office focused on technical aspects of armor.

---

54   Amiad Brezner, Origins of the Israeli Armored Corps (Tel Aviv: Ministry of Defense, 1995), pp. 52-5, 96-103. [Hebrew]

55   "Final Report of the Activation Team," 53-4. Similar ideas were presented after the committee's initial meeting in September 1949. "Protocol of the Armored Committee's meeting, convened 14 September 1949," I.D.F.A 154-488/1955. [Hebrew] The committee included most senior armored officers of the time: letter from Shalom Eshet to Chaim Laskov, 14 September, 1949, Op. cit.

56   Elron, "I.D.F. Infantry", p. 27.

When the office finally issued an operation-oriented manual devoted to infantry-armor cooperation (1951), it relegated the armor to assisting infantry attacks.[57] Similarly, a 1952 Doctrine Department pamphlet assigned operations deep behind enemy lines to a reinforced infantry brigade.[58]

Things changed in 1953, when the school for battalion commanders issued the Armor Pamphlet, mainly through the efforts of instructor Uri Ben-Ari.[59]

Already in its introductory remarks, the Armor Pamphlet set a new tone for the employment of armored forces. Rather than protection and firepower, the pamphlet emphasized armor's mobility, comparing it to the cavalry of old and advocating employing it en masse, instead of scattering it among infantry units.

Its objectives would lie deep behind enemy lines, and its missions would include breakthrough operations and exploitation of successful breakthroughs. According to the new approach, infantry was assigned to protect armor, and tanks were to deploy for counterattacks in the defense force rather than for bolstering its firepower. The manual emphasized the understanding that the tempo of armored operations required greater intelligence-gathering and dissemination capabilities and logistic self-sufficiency. For example, the mechanized infantry battalion would carry food and ammunition for two days and gasoline for ten hours. The Armour Pamphlet was innovative, an adaptation of early World War II German doctrine. To an extent, its principles guided armored operations during the Sinai War (29 October–5 November 1956) carried

---

57   Chief Armored Officer HQ, Tactics of Infantry and Armor deploying in Cooperation [May, 1951], pp. 6-7. I.D.F.A 27-1529/1952. [Hebrew]

58   Operations/Doctrine department, Long-Distance Penetration (1952), pp. 5-6. I.D.F.A 5-25/1954. [Hebrew]

59   Doctrine department/Pahad, The Armor Pamphlet 3 (1953), I.D.F.A 92-998/2000. [Hebrew] Uri Ben-Ari, The Battle for Modern Warfare: A Story of Israel's Armor (Tel Aviv: Ma'arachot and Ministry of Defense, 2005), pp. 47-9. [Hebrew] Few non-technical armored doctrinal pamphlets were published before 1956, and fewer still by the armored corps. See the correspondence on this issue in: I.D.F.A 43-160/1959. [Hebrew] See also: Brezner, Wild Broncos, pp. 277-8

out by the commander of the 7th Armored Brigade, Uri Ben-Ari.[60] Their performance completely altered the I.D.F.'s perspective on armor.

The merits of the battalion commanders' course notwithstanding, the I.D.F. also required a school for the education and training of staff officers and senior command positions. Hence, it decided to establish the Command and Staff College.[61]

In preparation, an activation team was assembled for the task (25 June 1952); it surveyed several similar programs abroad and also reviewed existing General Staff procedures and doctrine in order to adapt training manuals for the new school and improve I.D.F. doctrine.[62] The Command and Staff College was officially established two years later (1 June 1954).[63] The instructors composed a series of training manuals, the principal among which was *Staff Duties*.

The Doctrine Department issued a slightly revised edition of that manual two years later (1956). As it is identical to the Command and Staff College manual in most regards, the following discussion encompasses both.[64]

The commander alone is responsible for their unit, and their decisions form the axis of staff work. To be of assistance, a staff officer must do the following: anticipate the commander's intelligence and information

---

60   Armor Pamphlet, pp. 1-3, 5-7, 13-4, 20-4, 29, 44-9, 70-4.

61   According to Shai, the curriculum and structure of the college were modeled on the British Army's Command and Staff College at Camberley, even though the Israelis lacked the background gained from four years at a military academy. Hagai Golan ed., 50 Years of Jubilee to the I.D.F. Command and General Staff College (Tel Aviv: Ministry of Defense, 2004), p. 23. [Hebrew]

62   "Establishing the Command and Staff College Team, 25 June 1952," in Hagai Golan ed., 50 Years of Jubilee to the I.D.F. Command and General Staff College (Tel Aviv: Ministry of Defense, 2004), p. 13. [Hebrew]

63   " Command and Staff College- Warning Order," I.D.F.A 276-147/1961. [Hebrew] The battalion commanders' course may have been past its peak by then. Rafael (Raful) Eitan, A Soldier's Story (Tel Aviv: Ma'ariv, 1985), p. 52. [Hebrew]

64   Command and Staff College, Staff Duties (1954). [Hebrew] According to: Shai, Generalship, p. 128, a 1952 Staff Duties was based on a 1949 British manual, but I have been unable to locate that manual. The date of its publication is unknown, but can perhaps be gleaned from a reference to an Armored Brigade Combat Team (Op. cit, pp. 11, 145), which only existed in the I.D.F. from February 1954 through April 1955. Brezner, Wild Broncos, pp. 178-9, 185-7. This reference was changed to an armored brigade in the following edition: Matcal/Ahad, Field Manual 1-40-ה, Staff Duties (1956), p. 7. [Hebrew]

needs, filter the influx of information, adapt the commander's decision into orders, and supervise their execution.

The staff officer is free to voice any misgivings they may feel—until the decision is made, at which point they must keep their reservations to themselves.

A chief of staff or a brigade-level deputy commander coordinates the staff.[65] The primacy once accorded to the chief of operations is now conferred on the General Staff branch (Operations). In the new directorate, operations, intelligence, and training are accorded equal status.[66]

The chief of staff (a position that exists in the higher formations) enjoys unrestricted access to the commander. They advise the commander on all matters or arrange for others to do so, coordinate staff work, and assume command in the commander's absence. In this matter, the manuals retained the obfuscated division of labor within the staff between chief of staff and the deputy commander.[67]

Since the deputy commander is expected to perform their duties (such as commanding a secondary effort or coordinating the rear headquarters) and coordinate staff work, one of these roles is bound to be neglected in an emergency, to the detriment of the unit.

To maximize the staff's effectiveness, it's divided into shifts in order to allow a continuous 24-hour shift. In addition, to increase effectiveness, the manuals objected to splitting up the staff into administrative and tactical H.Q.s, due to the inherent resultant damage to staff work. The manual also improved on previous H.Q. designs.

First, the deputy commander is positioned in the tactical rather than the administrative section, perhaps in order to facilitate transfer of

---

65   Staff Duties (1954), pp. 4-6. Hence, the manual included an extensive discussion for securing this cooperation. For instance, see the discussion of meetings (Op. cit, pp. 49-56), or the operations journal (Op. cit, pp. 111-2).

66   Op. cit, p. 8. According to Shai, "Generalship", pp. 37-8, the Israelis adopted the British staff system and its procedures indiscriminately. However, as demonstrated by Doron Pinkas, The Origins of the Israeli General Staff, Seminar Paper (Pum, 2006), pp. 50-2. [Hebrew], the reality is more complex.

67   Staff Duties (1954), pp. 11-2.

command to themselves in a crisis.

Secondly, the size of the H.Q. was restricted to avoid the disadvantages inherent in inflated organizations. The command group also was diminished: when absent from the H.Q., the commander is to be accompanied by Operations (operations or intelligence), artillery, liaison, and communications specialists, with the chief of Operations remaining at the helm.[68]

The order process was designed as a recursive, three-tiered progression cycle: collation and processing of information and possible courses of action; decision; and, finally, monitoring of the execution. Though it could be shortened in emergencies, its basic structure should be maintained. Having said that, the discussion in the manual regarding the progression cycle ends immediately before the actual battle, thus indicating that staff work concludes prior to combat.

An additional problem related to the commander-staff relationship is that, while the manual stipulates that the commander devise the operational plan, the chief of staff is authorized to prepare the situation estimate. Considering that the situation estimate is the basis of any operational plan, one must be intimately acquainted with it in order to facilitate adaptation of the plan to changing circumstances during battle. Hence, this arrangement could prove to be an impediment.[69]

Following collation—a comprehensive analysis of all available courses of action, which will facilitate deviations from the plan at a later stage—a situation estimate is prepared.

On its basis, the commander formulates the operational plan, which includes intent, an outline of the method, a basic timetable, and a general force composition. The details are worked out by the staff, approved by the commander, and then distributed. As the process described is essentially cognitive—depending on inductive and deductive reasoning—it is

---

68   Staff Duties (1954), p. 144; Staff Duties (1956), p. 122. Regarding the consequences of overcrowding HQs see: Shai, Generalship, pp. 11-2.

69   Op. cit, pp. 15-6, 28, 109.

influenced by a constant set of elements. Hence, lists of such constants can save time and ensure that important factors not be forgotten. For the same reason, situation estimates should be written, at least as notes, to ensure that relevant factors are given due consideration.

The situation estimate includes an objective (based on the mission assigned by higher command) and an analysis of force correlation, terrain, time and space, logistics, security, climate and communications. Once all possible courses of action have been assessed, the most probable one is identified, the commander formulates their operational plan around it, which comprises intent, subunit missions, force configuration, estimated timetables, and phases of the operation.[70]

Once the operational plan is drawn, the second phase begins: the issuance of orders. Whether detailed or general, orders should include a precise and concise "intent" paragraph penned by the commander to guide subordinates in adapting their plans.

This provision is soon diluted when the chief of staff is granted authority to write that paragraph, thereby effectively shifting the axis of command and control from commander to staff. Moreover, the Operations officer is authorized to approve the final version of the orders in place of the commander prior to distribution.[71]

The significance of this point cannot be overstated: the written word is the commander's best means of ensuring that their intentions are conveyed to subordinates. If even one word is misunderstood, the best of plans might become unhinged. The axis was somewhat shifted back to the commander, as they were required to approve the "intent" paragraph.[72]

If circumstances necessitated the use of spoken orders, the commander was required to meet their subordinates in their units, rather than demand they arrive at H.Q.[73] In order to monitor the plan's execu-

70   Op. cit, pp. 15-9, 23-8, 30-8, 109.
71   Op. cit, pp. 62, 70, 80. Later, the manual stipulated that the Operations officer receive the "intent" paragraph from the commander, "word for word". Op. cit, p. 79.
72   Staff Duties (1956), pp. 11-3, 30, 68-9.
73   Staff Duties (1954), pp. 71-2, 83-4.

tion (the third phase of the cycle), the commander would rely on reports by subordinates, inspections, or liaison officers.[74]

Staff Duties charted new territory for the Israelis, offering a systematic approach to staff work, including standardizing a "military administrative office,"[75] preparation of situation estimates, and formulation of operational plans based on thorough analysis and contingency planning.[76] From the perspective of mission command, such systemization had some negative effects: the chief of staff and the deputy commander were considered merely as advisors to the commander. They were excluded from the process of formulating the plan, thereby curtailing their ability to assume command in the commander's absence and pursue the initial plan. Moreover, both were supposed to function as chief of staff, the hard core of staff work. This ambiguity concerning the division of responsibilities was bound to lead to confused staff work in an emergency. The damage was even greater, since the weight of command and control essentially shifted from commander to staff. A final lacuna was the virtual termination of staff work at the onset of battle, though the staff's duties clearly do not end at that point.

While these agencies were developing and disseminating combat doctrine for battalion or brigade-level officers, a new strategic threat and a different type of war were emerging.

When the Israelis went on the offensive in April 1948, a great exodus of Arab-Palestinians began, and, by the end of the war, some 700,000 refugees had temporarily resettled in camps in neighboring countries.[77] Israel's

74   Op. cit, pp. 19-20, 93-102. However, liaison officers are often given other responsibilities by the HQ to which they are attached. Regarding these methods: Agranat Commission Report (1975), pp. 1275-83, 1300-5. [Hebrew]

75   *Staff Duties* (1954), pp. 115-42, 158-69, 170-89.

76   As a general rule, many of principles included were not implemented in the Sinai War. Operations/Doctrine department, "Letter to Commanders 6 - Staff Work, July 1957". [Hebrew]; Operations/Doctrine department, Letter to Commanders - Staff Duties II, undated', I.D.F.A 17-161/1959. [Hebrew]; Pum, "Lessons of Kadesh 4- Staff Work, Undated," I.D.F.A 16-161/1959.

77   For conflicting (documented) accounts of the birth of the refugee problem: Benny Morris, 1948: A History of the First Arab-Israeli War (New Haven: Yale University Press, 2008); Yoav Gelber, Palestine 1948: War, Escape And The Emergence Of The Palestinian Refugee problem (Brighton:

refusal to accept the refugees back and the Arab states' unwillingness to naturalize them exacerbated their plight and caused dire economic straits. In growing numbers, they infiltrated Israel's mostly unsecured borders, robbing and attacking settlers in border settlements.

To stem the tide, the I.D.F. established a border police, strengthened the defenses of border settlements, and conducted retaliatory attacks against villages supporting the infiltrators. However, at least until the summer of 1953, a high proportion of the raids failed to reach their objectives, primarily as a result of poor navigational skills, deplorable physical fitness, and insufficient commitment to the mission.[78]

The infantry battalions were undermanned (officers, NCOs, and soldiers) and their recruits were of low caliber due to faulty prioritization criteria of manpower allocation. They were inundated with welfare problems and suffered from inconsistent training, defective weapons and ammunition, a shortage of firing ranges, instruction materials, and more.

To an extent, the problem resulted from the lack of a guiding hand equivalent to a ground forces command. During the post-war reorganization, it was decided that the General Staff would exercise operational and professional control over the ground forces (comprising mainly infantry) and operational control over naval and aerial forces.

By 1952, it was clear that this arrangement was unsatisfactory. Because of its prohibitive cost, the I.D.F. could not establish a new command for the infantry, so it was decided instead to form an Infantry Department within the Doctrine Department (8 February).[79]

Sussex Academic Press, 2006).

78   See Minister of Defense Lavon's comments in: "General Staff Meeting, 27.7.1954," I.D.F.A 7-636/1956. [Hebrew] On the reprisals see: Zaki Shalom, "Strategy Debated: Arab Infiltration and Israeli Retaliation Policy in the Early 1950's", Studies in Zionism, The Yishuv and the State of Israel 1 (1991), pp. 141-69. [Hebrew]; Shimon Golan, Hot Border, Cold War (Tel Aviv: Ministry of Defense, 2000), pp. 246-342. [Hebrew]; David Tal, Israel's Day-to-Day Security Conception: Its origin and development, 1949-1956 (Beersheba: Beersheba University Press, 1993). [Hebrew]

79   For other changes introduced at the time see: Martin van Creveld, The Sword and the Olive: A Critical History of the Israeli Defense Force (New York: Public Affairs, 1998), pp. 130-5; Edward Luttwak and Daniel Horowitz, The Israeli Army (New York: Harper & Row, 1975), pp. 105-11. Some of the changes were organizational, such as the establishment of Ugdot or the cancellation of the Southern Command (for fiscal reasons). Amiram Oren, "The Regional Command as a Framework

Its chief was named Deputy Head of Doctrine Department for Infantry with the rank of colonel, rather than lieutenant colonel, as was common among Doctrine Department branch heads.[80]

The Infantry Department

The new infantry department (March 1953) was headed by Yehuda L. Wallach, former commander of the school for battalion commanders and a future Ugda commander (division-size ad hoc formation) in the Sinai War.[81]

Chronically understaffed, the infantry department nevertheless boasted several notable individuals including Israel Tal, Meir Pilavski (March 21), and Rehavam Ze'evi (3 September), who, along with several others, would work long and hard to correct and improve the infantry's combat readiness.[82] When Yitzhak Rabin took over Instruction (December 1953),[83] he engineered a permanent redistribution of instruction functions among three departments: infantry (Yehuda L. Wallach, followed by Zvi Zamir [November 1955]), combat doctrine (David [Dado] Elazar), and training (Mattityahu Peled).[84]

As noted earlier, the Doctrine Department of the 1950's, included

for the Activation of the Ground Forces in the I.D.F., 1948-1956.," Studies in Israeli Society and Modern Jewish Society 14 (2004), pp. 452-3. [Hebrew]

80   The I.D.F. had even recently (November, 1952) canceled a front command (Eighth Command). Deliberations concerning the new department can be found in: Operations/Mivtzaim, Operations, I.D.F.A 112-11/1955. [Hebrew] Heads of the infantry branch formerly commanded a regular infantry brigade. Elron, "The Infantry", pp. 58-60. Initially, the new branch was seen as a temporary Doctrine department agency destined to develop into a full command. "Structure of the Instruction Department- Notes for Discussion, 18 June, 1953," I.D.F.A 29-433/1956. [Hebrew]

81   Some of the ideas that guided his work can be found in: Jehuda L. Wallach, "Cultivating the Infantryman", Ma'arachot 94 (July, 1955), pp. 28-42. [Hebrew]

82   A late 1953 report of 1st Brigade was considerably less bleak. "Summary of visit to 1st Brigade, 22 October 1953", I.D.F.A 29-433/1956. [Hebrew]

83   These are evident from the many weekly/monthly meetings held and topics covered in them. Office of the Head of Doctrine department, Doctrine department/Ahad Branch Meetings 1953-1954, I.D.F.A 29-433/1956. [Hebrew]

84   Operations/Matam, Ahad, Pahatz and schools, I.D.F.A 21-639/1956. [Hebrew] When he left, a few months before the war in Sinai (November 1956), Instruction became a department once again, forcing changes in its organizational chart, which stabilized only in 1957 after the war. Office of the Head of Operations, Manpower transfers, Establishment and Dissolution of Units, January-November 1957, I.D.F.A 909-1034/1965. [Hebrew]

many of Israel's most famous future generals. This is testament to the importance attributed to the department during those years.

These commanders' biographies and autobiographies tend to emphasize the operational chapters of their military service, allotting only a brief reference to their tenure in instruction.[85] Largely, they were "doers," combat commanders who were required to spend some time behind a desk, very often against their will. Nevertheless, instruction personnel published more articles in *Ma'arachot* in 1955 than ever before.[86]

When a situation warranted new doctrinal responses, as was the case in 1956 when medical evacuation procedures required reexamination, instruction officers joined a retaliation raid. In that same month, the department lost one of its newest recruits on such a raid.[87] In late 1953 Meir Pa'il spent two months overhauling the curriculum of the Infantry NCO and Officer Training courses.[88]

Members of the department routinely were called upon to supervise the I.D.F.'s annual maneuvers. These were the same men who issued *The Infantry Platoon* (in 1956).[89] A long manual by Israeli standards, the 430-page *Infantry Platoon* used simple and direct language and syntax, covering a wide range of issues intended to demonstrate the broader framework within which the platoon operates.[90]

85   See for instance: Hanoch Bartov, Dado: 48 Years and 20 More Days (Tel Aviv: Ma'ariv, 1979), pp. 79-80. [Hebrew]; Yitzhak Rabin and Dov Goldstein, Service Diary (Tel Aviv: Ma'ariv, 1979), pp. 92-4. [Hebrew]; Ariel Sharon and David Schanoff, Warrior: An Autobiography (New York: Touchstone, 2001), pp. 180, 210. This is indicative of the Bitsuism ([doing], action-orientation) ethos prevalent in the I.D.F.. Shamir, pp. 89-90.

86   This was true until February 1956, after which there was a drastic decline in their involvement for a few months.

87   Major Moshe (Musa) Efron, a former company commander in the Paratroopers Battalion, was killed on 26 September 1956 in a retaliatory raid against the Jordanian Police station in Husan. "Ahad Monthly progress report, September 1956," I.D.F.A 63-159/1959. [Hebrew]

88   Torgan, "Training", pp. 162, 181. For his thoughts regarding the training of junior commanders: Meir Pilbeski (Pail), "Training Junior Commanders," Ma'arachot 94 (July, 1955), pp. 42-9. [Hebrew]

89   Matcal/Ahad, The Infantry Platoon (1956). [Hebrew]

90   Absent from this volume is the battle drill introduced by British army veterans during the 1948 war, which figured extensively in earlier manuals, but was discontinued during Rabin's tenure. "Yitzhak Rabin to Yisrael Tal, December 12th, 1954," in: Yemima Rosenthal, Israeli Prime Minister Yitzhak Rabin, 1974-1977, 1992-1995: Selected Documents, Volume 1: 1922-1967 (Jerusalem: Israel State Archives, 2005), pp. 153-5. [Hebrew]

Following a discussion of the infantry battalion's structure and weapons, the manual defined the platoon commander's duties, chief among which was instruction, and delved into the principles of military instruction and leadership.[91] It then offered guidelines (normally in the form of advantages and disadvantages) for matters ranging from sanitation and combat intelligence to combined arms[92] to situation estimates, orders, and battle procedures. For instance, while recognizing the subconscious (and idiosyncratic) nature of the situation estimate and cautioning against rigidity and dogma, the authors proposed to commit it to paper, even in shorthand, so as to ensure that all important factors were considered. This was especially crucial, as it formed the basis for the operational plan.[93]

Elaborating on the variables influencing the situation estimate (objective, force correlation, terrain, time and space, possible courses of action, and the operational plan), the manual added that it should be based on facts as well as conjecture; the variables should be assessed in light of their operational ramifications.

For example, when designing a defensive position against an enemy believed to employ armor, the commander might deploy anti-tank weapons/obstacles or mine possible routes of approach or might prepare an anti-tank ambush.[94] Even when discussing technical aspects, such as the rate of advance in administrative and tactical movements or night marches, the manual presented guidelines and highlighted problematic issues, rather than dictated procedure.[95]

The treatment of combat forms included the following: attack, defense, retreat, ambush, meeting engagement, raid, patrols, urban warfare, and mountain/forest warfare. Among the basic principles of warfare, the

91   Op. cit, pp. 26-41, 42-51.

92   Op. cit, pp. 58-72, 73-81, 87-9, 90-8, 100-19, 119-30. The armor is assigned a strictly supportive role.

93   Op. cit, pp. 135-7, 150.

94   The Infantry Platoon, pp. 136-44.

95   Op. cit, pp. 182-5, 191-5. One of these is the psychological factor. Op. cit, pp. 156-62.

manual included "movement" and—in the spirit of maneuver warfare—explained that it included both mobility and "movement intended to create a situation accruing an advantage over the enemy."[96]

When discussing the principles of war (maintenance of aim; development and maintenance of morale, initiative, security, surprise; concentration of effort; and economy of force, flexibility, cooperation, and sustainability), the manual again stipulated that they should be implemented to guide rather than dictate action.[97]

The first of these principles was maintenance of aim:

1. One must clearly see the objective dictated by the higher commander and endeavor to pursue it determinedly.

2. Every consideration and evaluation should be conducted only in light of the objective.

3. The objective should be explained to all those participating in the operation in order to facilitate their cooperation in the effort to pursue this objective.[98]

The Infantry Platoon centered on the deliberate attack (flanking and frontal) rather than on the meeting engagement. This was the case, despite the prevalence of meeting engagements and the opportunities they offered.

Based on fire and movement, an attack should be organized in depth and pursued with all available force. In all attacks, firepower should play a significant role, in that it inhibits enemy activity and allows the attacker to advance while in contact. The attacking force divided into the following elements: assault (primary element), holding (the platoon's heavy weapons), reserves (for maintaining the momentum, exploitation,

96 Op. cit, p. 216.

97 Op. cit, p. 220. The list, adopted from Britain rather than the US, represented Doctrine department thinking since 1953. "Training Instruction No. 6/3/53: The Principles of War," I.D.F.A 33-63/1955. [Hebrew] The previous list approved by Doctrine department had yet to include morale and sustainability. Operations/Doctrine department, Principles of War: Summary, December 1949. I.D.F.A 6-1291/1951. [Hebrew] Yitzhak Rabin's lecture on the principles of war, given as commander of the school for battalion commanders, was identical. Rosenthal, pp. 58-70.

98 Op. cit, p. 218.

replacements, and defense against counterattacks), security (from ground and air assault through observation), and deception (either by the primary force or by a dedicated element).[99]

Despite a penchant for flank attacks, the manual discusses the merits of frontal attacks, recognizing that frontal assault might be less costly in certain situations. For such an eventuality, the manual offers WWI German *Sturmtruppen*-like principles: small groups, advancing silently or under fire cover, and establishing points of entry for the remainder of the assaulting force.[100] Regarding meeting engagements, the manual provides several guidelines: maintain the initiative and stay the course despite diversions unless necessary (maintenance of aim), direct the smallest force to deal with the obstruction (concentration of force), and, finally, maintain speed.

In contrast to earlier manuals, *The Infantry Platoon* demonstrated a marked preference for night attacks, while still presenting the advantages and disadvantages of such an attack. For instance, though darkness complicates control and coordination as well as orientation, it offers concealment, facilitates surprise, requires less soldiers and resources, and allows for greater maneuverability.[101]

The envisioned attack relies on maneuver and subordinating firepower to movement, as firepower supports the advance, rather than winning the battle. It is noteworthy, however, that the discussion devoted to fire support was brief enough to indicate that it not be perceived as an integral component. Nevertheless, the proposed defense was not maneuver-orientated. Not only did the manual focus on the technical aspects of constructing a defensive position but also the short discussion advocating WWI German depth and firepower doctrine hardly mentioned counterattacks.[102]

99   Op. cit, pp. 224-6, 243.
100   Op. cit, pp. 234-5. Similarly, see the discussion of Infiltration in: Op. cit, pp. 261-6.
101   Op. cit, pp. 243, 247. 250-60. See also the discussion of the fortified position: Op. cit, pp. 266-280.
102   Op. cit, pp. 283, 333-4, 338. This reflects a basic conceptual divergence regarding the nature of defense: the I.D.F. seems to have perceived defense as a temporary measure supporting offensive

The panorama revealed to the platoon commander was extensive and included a full range of forms and modes of action. Attention to issues such as prisoners of war or the breadth and depth of the discussion of the forms of battle indicated that officers are regarded as not merely platoon commanders but also potential company and battalion commanders.

This was reminiscent of the spirit of the Palmach (Haganah assault companies) where each leader carried a "marshal's baton in his knapsack." In line with previous field manuals, *The Infantry Platoon* offered a descriptive, rather than prescriptive, doctrinal approach, even when discussing seemingly technical issues, such as choosing between landlines or wireless radio communications. The authors stressed repeatedly that the examples given should not be implemented blindly. The description of the process and execution of the situation estimate were also in line with previous manuals.

Strengthening the I.D.F.'s existing maneuver orientation, this manual distinguished between movement or mobility and maneuver. Artillery, which was limited in the I.D.F. at that period and for many years to come, was expected to suppress rather than destroy the enemy. The manual displayed a marked preference for night attacks and flanking maneuvers, offering guidelines again reminiscent of late WWI German doctrine. From the perspective of mission command, then, the Israelis had made considerable strides, developing a combat doctrine that was maneuver-oriented, flexible, and dynamic. Nevertheless, the staff's functioning and structure were less maneuver-oriented and less supportive of mission command.

The End of an Epoch

During the seven years under investigation (1949–1956), the Instruction Department/Directorate devoted considerable resources to developing and disseminating doctrine through the activation team, the

action elsewhere.

school for battalion commanders, the Command and Staff College, and, finally, the infantry department. The combat doctrine produced, the type of doctrine it advocated, the structure of command it depicted, and its attitude toward combined arms warfare were at times innovative and maneuver-oriented rather than attrition-oriented.

It gave preference to maneuver over fire and flank attacks over head-on collisions, decentralized rather than centralized command (although increasingly restricted by a cumbersome staff system and inherently restrictive checks and balances), and infantry-based semi-combined arms. Finally, yet importantly, almost throughout the whole period, I.D.F. doctrine offered analytical tools rather than dictating solutions to problems.[103]

While its sources of inspiration varied, the I.D.F. relied primarily on experience garnered on the battlefield and from the British and U.S. experience. Nevertheless, as the above analysis has demonstrated, it strongly resembled German teachings. Since the Israelis did not draw from the German experience directly or systematically,[104] it seems probable that similar strategic and cultural constraints led both military establishments along the same path. This hypothesis, however, requires further investigation.

The doctrine produced between 1948 and 1956 during the wars was translated, edited, or written primarily by experienced combat commanders ("doers"), many of whom had commanded troops in 1948 and would do so again in 1956 and later. These included the commander of the Southern Command (Asaf Simchoni), his Chief of Staff (Rehavam Ze'evi), both division commanders (Yehuda Wallach and Haim Laskov), six brigade commanders (Uri Ben-Ari, Haim Bar-Lev, Shmuel Galinka, Israel Tal, David Elazar, and Zvi Zamir), a deputy brigade commander

---

103   The analysis also refutes several tenets in: Gil-li Vardi, "'Pounding Their Feet': Israeli Military Culture as Reflected in Early I.D.F. Combat History", Journal of Strategic Studies 31:2 (2008), pp. 295-324.
104   Dov E. Glazer, "Conceptualizing Command and Operations in the Israeli Ground Forces, 1936-1966" (PhD Dissertation, Bar Ilan University, 2011), pp. 3-5.

(David Carmon), and one battalion commander (Meir Pail). While this group did not necessarily lead training in the I.D.F., it nevertheless reflected prevalent thinking.

It is interesting to note in this regard that, a few years later, Ariel Sharon joined (or was exiled to) Instruction. During his time there, doctrine was formulated for capturing fortified enemy positions. Sharon's 1967 battle plan for the capture of Abu Ageila reflected that doctrine in almost every respect.[105]

To what extent was the doctrine they developed manifested on the battlefields of the Sinai War? Did the strengths and weakness observed on the previous pages materialize?

The answer, of course, is multifaceted. The I.D.F.'s doctrinal approach from 1949 through 1956, again, was essentially descriptive rather than prescriptive, endeavoring to teach students how to think rather than what to do in specific situations. It is no wonder that different courses of action were pursued at different times.

Having said that, speed and maneuver clearly played a greater role than did firepower, as is evident for example from the experience of the 7th Brigade, which barely utilized its artillery battalion. Faulty ground-air cooperation procedures and technology severely restricted close air support. The preference for night attacks was demonstrated time and again during the campaign (not always successfully), as in the 37th and 10th Brigades's attack on Umm-Qatef, the 4th Brigade's attack on Quseima, and the 37th Brigade's attack toward el-Arish.

The preference for flanking attacks, not absolute at this point, was demonstrated in some cases— as in the 7th Brigade's penetration of Quseima and the 10th Brigade's failed attack on Umm-Qataf—but was absent in others—such as the 37th Brigade attack at Umm-Qataf.

---

105   Compare: George W. Gawrych, "Key to the Sinai the Battles for Abu Agelia in the 1956 and 1967 Arab Israeli Wars," Research Survey 7 (Fort Leavenworth: Combat Studies Institute, 1990), pp. 87-117; Matcal/Doctrine department, Field Manual 1-2, Combat Doctrine, Vol. 1 (1964), pp. 66-68. [Hebrew]

Regarding combined arms warfare, it should be noted that, while the 7th Brigade deployed balanced battalion combat teams (not always successfully), most other brigades did not. Following the breakthrough operations, brigade objectives were set far behind enemy lines in an attempt to collapse the Egyptian defensive positions in Sinai.[106]

Postwar lessons disseminated to the units revealed problematic aspects of commander-staff relationships, where some commanders never made it to the front and others hardly visited their headquarters. The terms "Chief of Staff" and "Deputy Commander" were used interchangeably, and the person occupying that position expected to lead a secondary effort or coordinate staff efforts. Reporting was insufficient: H.Q.s were left in the dark for too long and prohibitively inflated, and so on.[107] Indeed, these faulty procedures continued to plague the I.D.F. in decades to come.[108] Armored doctrine took a new course, no longer being required to travel on transports behind the advancing infantry but participating in—and even leading—breakthrough operations before aiming for the enemy rear.[109] Concomitantly, the armor took on a more active role within the ground forces, achieving seniority by the early 1960s and attracting the best and the brightest.

In the armored brigades, which were still fairly balanced combined-arms formations, more commanders adopted a flexible approach to command, which was reflected in the 1965 *Armored Warfare* pamphlet and then on the battlefields of the Six-Day War.

A famous quote by 1956 wartime Chief of Staff Moshe Dayan has him preferring to fight with "noble horses than to spur on the reluctant

---

106 Regarding the military operations see: Moti Golani, There Will Be War Next Summer…: The Road to the Sinai War, 1955-1956 (Tel Aviv, Ministry of Defense, 1997), volume 2. [Hebrew]

107 Operations/Doctrine department-Torat Hakrav VeHamilchama Branch, "Letter to Commanders 6: Staff Work [July, 1957]", Op. cit. [Hebrew]

108 Shai, "Generalship", pp. 5-21.

109 For instance: Operations/Doctrine department- Torat Hakrav VeHamilchama Branch, "Letter to Commanders [unknown]: Armor in the Attack [unknown]," I.D.F. A 17-161/1959. [Hebrew] See also Laskov's analysis of the war: Chaim Laskov, "Lessons of Kadesh [December, 1956?]," p. 7, I.D.F. A 177-1034/1965. [Hebrew]

bulls." Clearly extolling subordinate initiative, Dayan was, in fact, also criticizing some of those initiatives, which on more than one occasion negated not only his plans but also his intent. In light of Asaf Simchoni's decision 24-hours earlier to commit the 7th Brigade, or Ariel Sharon's decision to reconnoiter the Mitla Pass by force, one can argue the inherent wisdom of these choices that clearly subverted Dayan's intentions. Thus, while the Israeli system might have possessed other significant virtues, this lack of commitment to the commander's intent was injurious to mission command.

Figure 4: 202nd Paratroopers Battalion at the Mitla Pass, 1956

Clearly, some of the perquisites for mission command existed while others did not. Some of those continue to haunt the I.D.F today.

The move away from combined-arms formations reached its zenith in the years preceding the 1973 war, thus sabotaging combined-arms warfare with headquarters expanded to bursting point and the division of labor within the staff never fully addressed or rectified.

These flaws notwithstanding, the I.D.F. managed to create its own unique path within the social, political, economic, and strategic constraints surrounding it and succeeded in maintaining a hard-won edge over its enemies.

Dr. Dov Glazer is an Israeli military historian whose academic focus has been the development of German and Israeli combat doctrine and includes more diverse issues, such as the 16th century Anglo-Spanish intelligence war and the commemoration of fallen soldiers in Israeli settlements' names.

Part 3

Leaders Talk

Editor's Note

Brig. General Gideon Avidor

After the Sinai Campaign of 1956, the I.D.F. began constructing a modern standing army. The main objective was to transfer fighting to the enemy ground by means of armored attacks with air fire support. During the Sinai Campaign, the I.D.F. ground forces were commanded by mission-based division headquarters, which assigned manpower according to need. After that war, the army established permanent divisions with a fixed structure based on armor. This sometimes comprised three tank brigades and other times comprised two tank brigades and a mechanized brigade. In addition, there were Special Forces and independent infantry brigades. The infantry divisions were territorial formations in nature.

The concept of the maneuvering armored attack had an impact on command and control doctrine. Commanding large, mobile armored formations maneuvering in depth demanded commanders and fighters with high professional standards as well as a system to allow rapid decision-making processes and the ability to alter forces' deployment and operational plans by means of radio communication. The example of the German blitzkrieg was in the background at all times.

Senior and junior commanders were in contact by means of forward tactical command posts. All commanders up to and including the brigade level moved on tanks together with the troops responsible for the major effort. Division commanders moved with the tactical command posts, and communication was direct.

A battle doctrine developed based on technical training at the low tactical level, freeing the commander to deal with organizing and deploying forces as well as assessing and managing the fighting. Junior officers trained to make independent decisions depending on fighting conditions and the mission at hand. Commanders at all levels are connected to one radio command system, so that they all could hear the commander and follow the course of the battle. In every command car, there was an additional radio for tuning in on the higher ranks or nearby support teams.

Battle procedures were either long (twelve hours or more), medium (three to six hours), or brief (up to two hours). The preferred battle management model was direct orders, presented as either graphic orders or orders issued over the radio, with arrows marked on the map. The use of coded maps that created a common language and served as an aid to control. All these could be conducted over the radio, without the need for face-to-face communication among commanders.

During the Yom Kippur War, I served as a G3 officer at 252nd Division Headquarters. In the course of twenty-three days of fighting, not a single written command was issued. All the battles, including crossing the Suez Canal and advancing toward the city of Suez, were conducted by means of graphic orders or orders issued over the radio.

At all periods, mission command was the preferred command style, but reality made this difficult. Limited operations focused attention on subordinate headquarters and factors external to the actual mission, and their repercussions were immediately felt in both public opinion and policymaking. Thus, the higher echelons are reluctant to take risks, on the one hand, while having the time and the wherewithal to control the forces under their command, on the other hand. Such a situation was evident in the fighting in built-up areas that the I.D.F. was involved in, starting with the First Lebanon War of 1982.

Another example was the policing activity carried out in Judea, Samaria, and the Gaza Strip. Fighting in those areas took place in contained spaces against localized objectives.

In such incidents, maximum reliance on accurate intelligence, involvement of civilian populations, and constant media presence on both sides of the battlefield introduced a consideration of "avoiding non-military complications" into battle planning and management. Headquarters employed legal advisors, army spokespersons accompanied the forces, and commanders limited freedom of decision, even in areas that are not directly involved in the fighting. Thus, in such cases, detailed command is the easiest way to avoid misfortune. In an unresolved war, the enemy remains active and aggressive; after its conclusion, they are a partner (unfriendly) to the war results, and the media declares both sides the winner. Two striking examples of this are Hezbollah in Lebanon and Hamas in Gaza.

The I.D.F. has another contingency that is not present in other armies, namely, integrating reserves into limited operations. Reserve forces participate in training or operations for brief periods and change over at three-day intervals. They lack the sectorial expertise and accumulated knowledge necessary for active duty in "between the wars" missions, so local command centers must guide, support, and supervise their activities and outcomes. Again, detailed command is the safest way, whereas in conventional fighting situations, reserve units can function under mission command without difficulty.

Israel's unique situation dictates that the transition from low-grade to conventional battles might be quick and sudden. This is not easy and the costs of transferring from a routine of sporadic fighting to full-scale warfare demands time and adjustment, as is attested to by the outcomes in the first days of fighting. This was the case in the 1973 Yom Kippur War and the 2006 Second Lebanon War. In the 1967 Six-Day War, the I.D.F. had only three weeks to prepare for war in a sharp, rapid transition.

The Six-Day War was the I.D.F.'s ultimate test as a modern standing army. It ended in decisive victory for Israel, and the army worked diligently toward applying the insights gained and lessons learned toward the future. The commanders in that war were promoted to the general staff and prepared the forces for war based on their experience and wisdom. The outcomes indicated to them that the army was on the right path. For the I.D.F., the Six-Day War was the model for approaches and doctrines for the army of the 70's.

Unfortunately, war is not one-sided, and the enemy learned its lessons from the self-same events. This came to expression in the 1973 Yom Kippur War, when the I.D.F. found itself in a gap between what it had prepared itself for and the reality. The regular force that was assigned to block Egyptian and Syrian attacks found itself in the minority. Until the high command recovered and called up reserves—and the Egyptian and Syrian attacks stopped—the regular ground and air forces had to bear the brunt of battle. Reserves were sent into battle as they arrived and functioned up to the division level according to mission command, and that was the key to successfully arresting the attacks and leading the army to victory.

The element of surprise, mistakes in deploying forces toward the war, and inferior numbers until the force recuperated its strength all placed the burden of hand-to-hand fighting on the lower tactical ranks, which were left to face the enemy in the first days of fighting.

These were tactical forces on the platoon, company, and battalion levels, which had to find solutions for unplanned situations without the support of the higher ranks. The sources of those small forces' strength were leadership, motivation, and professionalism. Commanders took the initiative, leading their forces into battle without waiting for orders or backup. Every one of them fought to the best of their ability, and their soldiers fought alongside them. Mission command was the leadership model during the first three days of fighting up to the brigade and even to the divisional level.

The fighting spirit of commanders in the 188th Brigade's platoons and companies on the Golan Heights, each of which fought alone against the onslaught of Syrian tank regiments pushing westwards, brought the enemy to a standstill.

They continued fighting after the brigade commander, his aide-de-camp, and the operation officer were all killed when their tank was hit and brigade headquarters ceased to function. The same spirit was present among the regular battalions fighting on the Suez Canal front. Without mission command, the outcome might have been very different.

After the Yom Kippur War, circumstances changed, and the I.D.F. along with them. The First Lebanon War of 1982 no longer involved maneuvering against a regular army. There was no longer an existential threat against the state of Israel. Fighting tactics changed, but commanders and command systems remained unchanged. More significant changes took place in the late 1980s. There was no longer the threat of attack by regular national armies, but fighting against terrorist organizations was on the rise, while significant technological innovations and digital systems were introduced into weaponry and command and control systems.

Fighting against terror in built-up areas among civilian populations forced commanders and command centers to pay more attention to the effect of their activities on local and international public opinion, furthered by the constant presence of digital media on the battlefield for purposes of reporting. Both of these factors had a dramatic impact on the commanders and their approach.

The span of time between incidents shortened considerably, and the ability of the higher echelons to monitor every action from any location imparted a sense that "Big Brother" was watching. Indeed, the higher ranks realized that they could exert complete control over junior commanders, who felt threatened by this. The Second Lebanon War of 2006 was a prime example: brigade and divisional commanders chose to manage the fighting from rear headquarters by computer and hesitated to initiate attack strategies for fear of negative public opinion.

In the First Lebanon War in 1982, there was a clash between the tactical level's expecting to act according to mission command and the divisional commanders' lack of decisiveness and clarity regarding planning and orders. Confronting an unconventional enemy, for which the I.D.F. was doctrinally unprepared, only made matters worse and added to the confusion.

The coming years saw low-intensity warfare against various organizations. The major threats to the state of Israel were internal terrorism and long-range firing on civilian settlements. The I.D.F. began focusing on protecting the borders with a range of technological developments and attacking the source of the firing, mainly from the air. Maneuvering became less important than firepower, greatly influencing leadership and command and control systems. In this type of warfare, technological becomes the decisive factor and activity centers around small frameworks; thus, the importance of Special Forces increases, and they are responsible for most of the ground fighting. Mission command was always popular in such frameworks, in which personal example and internal leadership constitute the dominant ideology.

In such circumstances, the importance of large formations lessens. Part of their effort is directed at supporting Special Forces, generally according to detailed command, as supporting forces cannot function according to mission command. A situation is created in which the dependence on technological means and the need for operations to be exact and closely vetted create a conflict of interest with the mission command approach, which, according to I.D.F.'s doctrine, is the preferred style at the junior command level.

Mission command is a culture and, as such, is not limited by hierarchy. It appears at every rank and is carried out both formally and informally among commanders. It does not challenge authority or responsibility. It is widely discussed throughout the I.D.F.; junior commanders ponder its significance and senior commanders act according to its tenets.

Subordinates are encouraged to express their own operational ideas and their superiors always weigh them up seriously, either accepting or rejecting them, but always giving encouragement for their initiative. There were, and still are, arguments concerning mission command which are influenced by differences of approach and personality, and it is not always very successful.

In this section, we have chosen to present a commander's insights and perspectives, each in their own specific historical context and experience. We picked senior leaders at the Star Officers level and field commanders from platoon leaders to division commanders.

The field commanders' thoughts are presented in chronological order, since thoughts derive from knowledge and experience; the last is much related to the author's personal experience. A common thread runs through all of them, but each has its own unique character and style.

1

A Quick Guide for the Junior Officer

Lt. General Haim Laskov

It is impossible to clearly define what is demanded of the commander and their leadership if we do not first visualize the "proper conditions" that are likely to be present on tomorrow's battlefield. We might say that, on tomorrow's battlefield, the same qualities that have been expected of the commander will continue, but even more so. Such fighting conditions demand performance according to plan—or constant adjustment of plans— but always with adherence to objectives. What will be demanded is active, agile leadership that can achieve real results in a short time. Fighting conditions will demand extreme daring and speed.

If future leadership is to reach the peak of its abilities, then the basic qualities demanded of each commander are:

- Physical fitness
- Intelligence and discernment
- A sense of movement and fast pace
- Strong nerves and composure
- Aggressiveness and daring

The lower ranks on the front lines, from the squadron commander up to the battalion commander, are taking on much greater importance. The key to this will be the commander's ability to trust their soldiers and inspire them to follow their leadership. In order to do this, the commander must know how to command.

One of the basic requirements of a commander is responsibility. It is possible to distinguish two areas of responsibility that have no clear boundaries between them: one stems from the will, strength, and orders of the higher-ranking commander responsible for a given unit, while the second stems from the will, strength, and orders of the commander of the unit themselves.

The positive behavior demanded of the commander is as follows:

- They must gladly accept responsibility and not avoid it.
- They must not hide behind the backs of the higher ranks.
- They must not merely repeat orders issued from above but must adapt them to the present situation.
- They must not look for scapegoats.

From the moment an officer enters their post until the moment they leave it, the commander must become accustomed to making decisions and carrying them out. In light of what is demanded of the commander, through discussion, training, and solving tactical problems, every course, every officer's training college, and every commander responsible for the development of their junior officers must strive to inculcate the following qualities into the future and beginning commander:

- The ability to make decisions and give orders.
- To remember that a command issued in person has the same validity as one given in writing (and that the time element might even give it priority!).
- Decisiveness, forcefulness, and the ability to concentrate efforts to carry out a particular mission.
- Not to criticize or analyze plans and maneuvers while in action, but to strive to achieve a decisive solution and maintain a fast pace in performance.
- To master the techniques of the art of command, not in theory, but in practice.
- Loyalty to those above (i.e., the willingness to take full responsibility for carrying out the orders one receives).
- Loyalty to those below (i.e., giving full support to subordinates).

- A large degree of self-control and avoidance of exaggerated ambition.

How can these be achieved?

1. When the situation is complex and deteriorating quickly, simplify it in order to see what is important.
2. Make sure to clearly explain to soldiers what is expected of them.
3. Develop bravery, not only in face of the enemy.
4. Consider the issues and learn your profession and your role.
5. Concentrate on achieving goals and be daring in order to achieve them.

LTG. Haim Laskov joined the Haganah as a teenager and served in various units, including Orde Wingate's Special Night Squads. In 1940, Laskov joined the British Army so that he could participate in World War II. He served in various capacities, eventually reaching the rank of Major. When Israel's War of Independence erupted in 1948, Laskov assumed responsibility for preparing the framework in which new recruits would be trained. He organized the first officers' course; in that war, he commanded battalion and brigade in various operations.

Although he had never been a pilot, Laskov was appointed commander of the Israeli Air Force in 1951. Upon completing his tenure in 1953, Laskov left the army to study philosophy, economics, and political science (PPE) in the United Kingdom.

In 1955, he returned to Israel. During the 1956 Sinai Campaign, he commanded the 77th Division, which operated on the Rafah-el-Arish-Kantara front.

In 1958, Laskov was appointed Chief of General Staff, replacing Moshe Dayan.

Laskov resigned his position of Chief of General Staff in 1961.

2 Leading and Command

Maj. General Israel Tal

If up to the Yom Kippur War there was one common goal that was acceptable to all, a kind of "Archimedes' Law" set in place after that war, it would be the weight of the individual's motivation and willingness to take part in the communal effort. Beforehand, we depended on unity of purpose. Leadership today must rely more on the preparedness of the individual, but this is not sufficient alone. A number of operative steps must be taken—even some very unpopular ones. The new conditions demand more reliance on the formal side of the organization, on formal discipline. Attention must be given to developing proper frameworks in accordance with this.

Against Popular Leadership

We have always claimed that techniques and training are basic schemes that can be applied according to need, as deemed necessary by commanders. At the same time, we must ensure that the application does not deviate from formal frameworks and the basic disciplinary principle "that an order is an order by virtue of being an order!" Schemes and discipline are inextricably woven together, and one validates the other. Professional authority must be enhanced, and the commander must be an excellent top-grade professional.

Another vital element is increasing moral authority that is derived from personal example in all fields, while placing special emphasis on interpersonal relations and the approach to other people in general.

Excessive boasting among corps should be replaced by pride in one's corps.

In conclusion, in the reality following the Yom Kippur War, it is vital to behave in a decent manner, not according to consensus. Those who practice correct leadership are liable to feel that their subordinates are ungrateful, but—in real time—they will respond to cheap, opportunistic, popular leadership with the contempt it deserves.

[Aleksandr] Solzhenitsyn wrote in his book, *August 1914:*

There is no end to the suffering of an exceptionally talented officer. The army enthusiastically surrenders to talent when it already holds the baton of power, but the more it is attracted to that baton, the more it will suffer endless blows.

Parallel to this, those of you who choose the right path are liable to suffer, but, at the decisive moment, your subordinates will reward you. They will follow you and will not disappoint you. That who chooses the easy, deceitful path will win sympathy on the surface but will quickly find this turning into contempt. Do not go for cheap, fictitious leadership. It must be spontaneous and according to the best conscience of every one of you.

Maj. General Israel Tal (Talik) was born in Zephath in 1924. Tal began his military service with the British Army's Jewish Brigade, serving in Italy during World War II. In the Independence War, he commanded a platoon in the Givati Brigade. Following that war, he served in command and instruction roles.

In 1954, he studied at the Senior Officers School in Britain. Tal served as an infantry brigade commander during the 1956 Sinai War. From December 1956 to April 1959, he served as deputy armored corps commander. Later, he served as assistant head of the Operation Directorate of the General Staff and as Chief of Staff at the Northern Command.

In 1964–1969, Tal served as chief of the Armored Corps. He played a major role in the "War on Water" with Syria in the north of the country.

In the Six-Day War, Tal commanded an armored division that fought in northern Sinai.

In 1972, Tal was appointed head of the Operations Branch in the General Staff, and, in 1973, he became deputy Chief of Staff, a role in which he served during the Yom Kippur war.

In 1974, Tal retired from the I.D.F. and established the Institute for National Security Studies at the Tel-Aviv University.

Tal continued to be involved, actually carrying overall responsibility, in the development of the Merkava tank and is considered the "Father of the Merkava" and "Mr. Armored Corps."

General Israel Tal died on September 8, 2010. His Book *National Security: The Israeli Experience* was published by Praeger Security International in 2000.

3 Mission Command Culture

Maj. General Doron Rubin

The adoption or rejection of mission command reflects an army's culture, and the term "culture" here relates to hierarchy. The true test of an army is undoubtedly under wartime conditions, when junior commanders must make their way without any specific instructions. However, mission command is also present with those who are accustomed to a very different culture.

In peacetime, in training, in ongoing security activity and special operations, commanders are often reluctant to rely on subordinates to complete missions successfully. This is directly related to their being unprepared to suffer failure and criticism.

Although modern technology makes it possible to receive all information simultaneously, in order to exploit it to the full, information must be used wisely. This means that the commander does not need to know everything at all times but only what is required to enable them to make decisions. At times, the commander feels the need to step into the center of gravity and guide by their own opinions. This is a well-known phenomenon, but mission command demands that leaders resist this urge and limit their involvement to the moment when they must make a decision. This is generally influential in critical junctures that can affect the entire operation and so demands clear and decisive intervention, like a surgeon performing an operation to save a patient's life.

Undoubtedly, mission command demands patience and the ability to rely on junior officers' abilities to carry out missions; it is in the army's best interest to develop an atmosphere encouraging such a command style. If they, at times, encounter a commander giving their juniors freedom of action and are criticized for being shallow and reluctant to deal with details—and such opinions—then it will be hard for the junior commander to follow this doctrine.

However, an in-depth examination of this command style makes it clear that, in fact, mission command embraces all the "details" necessary to support and encourage junior officers.

The Importance for the I.D.F.

In times of war, systems often break down, and it is impossible to approve junior commanders' plans or provide them with specific commands; they might also find themselves in rapidly changing conditions. In such cases, junior officers who are not trained for such situations will find it difficult to overcome them.

In routine security operations, commanders' sensitivity levels are high, thus, they tend to intervene actively, even when doing so is superfluous. The drawbacks of such behavior are obvious to everyone.

First, it places unnecessary pressure on junior commanders, as well as battalion or brigade commanders, who also complain about senior commanders' intervention to the point of loss of judgment. Secondly, such a command style educates junior officers not to take responsibility. Thirdly, and most seriously, they do not become accustomed to commanding independently, which will be crucial in face of the greatest test of all: war.

Of course, there is a price for giving juniors freedom: the danger of failure. However, this price must be paid, since officers are only able to prepare themselves for action in wartime by means of events in routine times. It is thus necessary to weigh the price that senior commanders are liable to pay in the future against the price of not allowing juniors to make mistakes, which will be far costlier. Encouraging mission command in the army should be done on two levels: formal training and training through routine activity.

The quandary between reality and doctrine recently came up in an article by a platoon commander who found it difficult to apply mission command in the field. The idea that the theory is correct but that reality is something quite different demands explanation. There are two fundamental ways of dealing with it. The first is to abandon the idea and return to a familiar hierarchical rut, and the second is to adhere to it as closely as possible with certain limitations. Such a discussion constantly preoccupies every I.D.F. commander, since culture dictates one thing and life and the field present a very different picture. In our army, even when command strictly is imposed from above, mission command is always in the background and is the subject of endless controversy.

Maj. General Rubin was drafted into the Israel Defense Forces in 1963. In 1964, he became an infantry officer and returned to the Paratroopers Brigade as a platoon leader. In the Six-Day War, he served as a company commander and fought in the Gaza Strip. During the Yom Kippur War, Rubin commanded the 202 Paratroop Battalion. Later on, he commanded the Israeli Defense Forces' Officer Candidate School. His next command was as commander of the 35th Paratroopers Brigade. In the 1982 Lebanon War, he led the 500th Brigade. Later on, he commanded the 162nd Division. Afterwards, he was appointed as the head of the Israel Defense Forces' Instruction and Doctrine Directorate.

4 Reflections of a Staff Officer[1]

Lt. General Motta Gur

I would like to express a few opinions about the staff officer. This is one of those topics that cannot [be] measured or estimated; there are no measures or tests that can do so. It might [be] observed in the officer's image but not necessarily in their daily functioning. Thus, this situation appears to resemble the proverbial bread [that] cast upon the waters. If we take those words literally, anybody can extend their hand to the bread floating on the water and take a bite, but if we accept the biblical meaning, the main beneficiary will be the I.D.F. itself in years to come.

The staff officer's job is easy to learn, practice, and apply. However, their spirit, work ethic, and behavior cannot [be] learned by means of exercises, as they [are] based on values that can only [be] inculcated through an extended learning process.

There is a well-known saying: "The army marches on its stomach." Since the importance of the stomach is already familiar to everyone and armies know, more or less, how to satisfy it, and since the army's image since Napoleon has changed, it might be appropriate to consider another saying: "The army marches on its values." In the present article, we will consider values in the context of the staff officer.

The Staff Officer as Compared to the Commander

It is impossible to discuss the staff officer without viewing them in their natural habitat—together with their commander. Therefore, the

1   Reprinted with permission from *Ma'arachot*, Vol. 342, 1995, pp. 2–7.

major part of the following analysis will relate to the relationship between the two: the staff officer's status regarding their commander and the higher ranks as represented by the commander. In the present article, I will not deal with the commander, as many have already done so. There are many different ways to deal with the commander-leader, but it seems to me that it is not far from the truth to state that there is a consensus regarding their image and behavior.

The situation is different regarding the staff officer: in different staffs, with different commanders, in different units, and even in different situations, the staff officer is required to behave differently. The fact that they are not the center of gravity apparently prevents him from creating an independent entity. They are always dependent, drawn by gravity to the sun—the commander—or repelled by it. They go around in circles, at times in the light and at times in the shadows. There are no definitive prototypes regarding them.

The Command Branch and the Staff Branch

The question may [be] asked if there are two breeds of officers in the army: the commander on the one hand and the staff officer on the other. Does each require a different character and essence (as the officer must decide during their career whether they are suitable for command or staff work)? We are familiar with examples of talented officers in the world's armies who were obliged to serve as staff officers throughout a particular war, waiving the glory of commanding on the field of battle. How did this come about?

An important I.D.F. institution is rotation, by which the officer occasionally [is] transferred from staff to command and vice versa. Does this mean that any officer is capable of being either a commander or a staff officer? It appears that this is not the case. I am familiar with two major types of officers. For the first type of officer, the main tree they climb is the command tree. To do so, they are occasionally aided by side branches, by means of which they climb in the form of brief stints as a staff officer, whose main purpose is to aid them in advancing as a commander. Conversely, there are officers who mainly serve as part of the staff and this constitutes their main tree. At times, they leave the staff and take up command posts, but for them, these are the side branches. They do so in order to view the other side of the coin, broaden their perspective, see things through the commander's eyes, and attempt to delve into the problems that face the person that the staff [is] meant to serve. These experiences only serve to help this type of officer to perform more fully their main role in the military hierarchy—as a staff officer.

The Staff Officer's Qualifications as Opposed to those of the Commander

Above all, we demand of the commander that they be successful on the battlefield. As long as they maintain reasonable human relations, we do not analyze and probe into the circumstances under which they achieved success. (We leave such analyses to writers of memoirs and military historians.) Alongside success on the battlefield, we want the commander to be decisive and practical. To sum up, the commander is required to make decisions, to act accordingly, and mainly, to succeed. We expect the staff officer to possess different qualities:

- Independent ideas and the courage to express them;
- Analytical thinking skills;
- The ability to clearly and convincingly present facts, data, conclusions, and recommendations;
- Loyalty;
- Willingness to be the person behind the scenes and "step into the limelight" only when necessary.

Discipline and Adaptation

First, I will indicate two assumptions on which to base my observations. The first is that I.D.F. officers [are] disciplined. They know how to give orders and accept them. Every officer knows that without absolute discipline and basic loyalty to their commander and the matter at hand, it is impossible to carry out orders or perform missions. It is true that for every concept mentioned above (discipline, loyalty), it is possible (and desirable) to offer endless interpretations, but I am sure we can agree that I.D.F. officers are not lacking in discipline.

The second assumption is that an army officer's life is not always trouble-free. They have many worries and must take a stand every hour of every day regarding problems, whether big or small and whether individual or general. Furthermore, their life becomes even more complicated when they are promoted to a higher rank.

The army is not lacking in hard, impatient commanders who are convinced that they are right and who lack the patience to cooperate with their staff or work as part of a team. Their path is clear and they know exactly what they want and how to achieve it. A commander such as this expects their aides not only not to bother them, but also to obtain all the resources necessary for them to achieve their objectives. It is often difficult to work with such commanders or confront them with opposing opinions; at times, it might even be risky to do so.

I am well aware of this situation and my further comments consider it. I am also not ignorant of a widespread assumption among officers regarding career advancement; I am referring to a situation in which every officer must select a "horse" which they can ride toward promotion. The higher the rank and position of this "horse," the better. If one wishes to advance rapidly, the "horse" must [be] treated with delicacy and caution. One must not fight with it or anger it, but rather pacify it with sugar cubes.

The Staff Officer's Struggle

It may [be] assumed that we [are] indeed disciplined, but army life is difficult and filled with obstacles, such as difficult commanders that cannot

[be] reasoned with. Our personal careers are very important to us, so at times in order to climb the ladder, we are willing to put aside principles and mount "horses." So what should we do? If life is really so hard and it is useless to argue with the commander, since they will not listen anyway and we are beating our heads against a wall, what can be done?

We Must Put Up a Fight

As with any problem, there are a few possible solutions, but I will only mention two of them. One way is to surrender, accept the situation, knuckle under, and attempt to regain equilibrium. Everybody can discover within themselves psychological defense mechanisms to justify their position and worldview. They might rely on examples from personal experience or the hallowed halls of history—both general and military. They will probably also find comfort in "life truths."

The other possibility is to "stand and fight." Do not underestimate the "enemy"—especially if they are an Israeli commander. However, do not let them chase you away either. Quite the opposite; they are someone you can reason with—you just need the courage to do so. Furthermore, the Israeli commander, who prefers to fight wars that are worthy of their name, will probably enjoy the confrontation.

Such a struggle is often complex, hard, and painful, but it is the only way officers and armies may [be] led to achievements and advancement, rather than to stagnation and moral corruption.

The Nature of the Struggle

Let us clarify exactly against what we are "fighting." Against what must the staff officer muster all their mental strength and go out on the offensive?

- Against routine;
- Against widely accepted conventions;

- Against the higher echelons' evaluations and assumptions;
- Against other staff officers' evaluations and assumptions;
- Against his commander, when necessary.

However, how should this struggle [be] conducted?

Against Routine and Convention

We can struggle against routine and convention by a periodic re-examination of the facts in front of us in [as] objective a manner as possible. We must seek innovative solutions for new problems, which will still [be] base[d] on existing doctrines and methods, but will not be limited by them. I purposely said new solutions, not original solutions. Originality is a God-given gift, which [is] not bestowed on everybody, especially when it [is] combined with deep commitment. New solutions are not contingent on originality, but are the result of breaking free from convention, relating to new problems, and changing data with an open mind. An orderly, methodical, consistent, and ongoing analysis will necessarily reveal new horizons. People like to rest on their laurels; it is easy for them to rely on consensual, accepted, and tried-and-true doctrines. Of course, I do not claim that innovation is appropriate in all conditions and at any cost, but I [am] perplexed by the blind acceptance of convention—and I suggest fighting against it.

Every army has a master plan regarding how to behave in time of war. On numerous occasions in military history, armies remained loyal to those fundamentals in face of hostilities.

Their assumption was that decisions formulated in peacetime, through deep analysis and profound thinking, would always be preferable to decisions based on hasty estimations of the situation and taken under pressure, as well as due to the fact that armies were organized, trained, and accustomed to functioning according to doctrines formulated in peacetime and believed that these could not be changed in the midst of fighting a war.

France's resounding defeat in the two world wars was a result of such an attitude. Obviously, we cannot allow ourselves to enter into a similar situation. Thus, it is obligatory that staff officers put aside convenience, examine every accepted convention, and, when necessary, combat them in an open, unrelenting campaign.

The modern army has at its disposal huge amounts of complex resources; the professional staff officer must search for new ways to exploit them. Technological innovations; transformations in the battle order and the enemy's fighting methods; as well as preparations for a specific, concrete war, with all its advantages and disadvantages, must all spur the professional staff officer to provide original, daring, and unique answers for each scenario. They must shake off their former ideas, assumptions, and suggestions, which in the past were a touchstone for their commander, and view current reality with fresh eyes.

The Staff Officer and the Higher Echelons

Regarding the "war" against the higher ranks, every officer—whether they are a commander or a staff officer—must view themselves as standing before the judgment of history. The commands neither they are given, nor the principles that guided them, will stand in their defense. The officer alone determines their path on their own personal responsibility, and they must be prepared to accept the verdict.

The needs and preoccupations of the higher ranks do not always correspond to those of subordinates; in fact, the assumptions and conclusions of the senior and the junior ranks often [are] diametrically opposed. Such a conflict among ranks will often occur when the general command directing the war issues orders as a whole that might be an appropriate response to the enemy, but could endanger a subordinate force. It is the staff officer's responsibility to identify such a situation and find ways of dealing with it.

Every organization determines its own self-image, interests, responsibilities, and objectives. The staff officer is obliged to check periodically whether their daily activities correspond to the requirements of the higher echelons—otherwise "flare-ups" might develop between the two levels that are liable to affect directly and adversely their unit. However, the staff officer cannot hide in the shadow of the higher levels' commands; they should relate to them as a challenge rather than as a way of shirking responsibility.

The Staff Officer and Their Colleagues

It is surprising to observe the ease and speed with which a particular staff can often arrive at identical thought patterns and work methods. This is due to a social atmosphere influenced by the commander and by officers with strong, influential personalities; obviously people like to feel a sense of belonging. Nevertheless, it is unacceptable for the staff officer to [be] carr[ied] along by the consensus without making their own unique contribution.

Therefore, the staff officer must sometimes stop and examine if the staff's communal spirit is the same as their own or if they disagree with some of its assumptions and solutions. The professional staff officer must make a personal assessment, not of their own career (although that is also important), but of the issue at hand. As the representative of a particular corps, they must produce a reliable professional assessment, according to a predetermined standard. As its representative, their professionalism and inner truth must be the basis for the assessment, which will later be compared with those of their Colonel leagues. Thus, in a harmonious staff of professional officers, each individual must consider themselves to [be] an integral part of the matter at hand. In the course of discussion and debate among staff members, a complete picture will emerge. Thus, the staff officer should not be enmeshed by solutions and procedures suggested by the commander or others. They must follow their own path—regarding the correct and optimal use of the resources that they represent, but also regarding the solution to the entire problem. Such an internal staff-based negotiation will remove any possibility of missing good ideas and will enable comprehensive, in-depth treatment of the subject under discussion.

The Staff Officer and the Commander

Here we arrive at the most sensitive issue of all—tantamount to splitting the atom—which is liable without proper supervision and control to generate enough negative energy to place the entire army in jeopardy, namely, the struggle with the commander. Careless, ill-considered behavior in this regard is liable to disrupt the army's foundations and result in a dangerous loss of discipline.

However, although the relationship between commander and staff officer is such an explosive topic, it is also a crucial one, so we will give it careful consideration. The commander's motivations are strong and vigorous and their attention [is] primarily directed at the practical aspects of the forces they command. It is both natural and healthy that they [be] focused first on their own behavior and that of their formation. The staff officer's role is to ensure that the commander makes decisions and acts according to the intentions of the higher echelon and remains faithful to them.

Loyalty

The issue of the officer's loyalty to their commander or mission is theoretically not very complex. Military history can supply us with countless positive examples of officers who deviated from loyalty to their commanders when the latter did not act according to what [was] judged as loyalty to the nation, the state, and human values. We have only to consider the example of General de Gaulle's rebellion against his government and his army in their surrender to the Nazi conqueror. In addition, the Nazis themselves supply us with a horrific example of an entire army that remained loyal to its commanders, thus ignoring any semblance of military ethics or basic human values.

I am not aware of a single example, from the I.D.F.'s establishment until the present day, when the question of loyalty [was] raised as a severe problem. I would like to hope that this will never happen, but every officer must be on their guard and view themselves as personally responsible for perpetuating this healthy and desirable situation.

The Staff Officer and Their Country

I assume that the senior staff officer, as a staff college graduate who is experienced in military matters, is capable of expressing an opinion regarding any problem that arises at the level at which they serve. This is not only a privilege; the staff officer has the obligation to express their opinion (tactfully, wisely, and courageously) and defend it.

Does this mean that not only the commander, but the staff officer as well, must have their own "policy?" The answer is that at the highest levels—in the field division, general headquarters, and general staff—this must indeed be the case.

Before I enter this minefield, I wish to return to the basic premise upon which I based my remarks until this point: I.D.F. officers' discipline. In my entire career in the I.D.F., I have never encountered a lack of discipline—especially operational discipline. Therefore, I am not worried about damaging or disturbing any hierarchies.

In fact, I am working under an entirely different assumption: our officer class [is] often too disciplined. In fact, this situation might be much more comfortable than fighting for one's opinions. It is difficult to object to an army's officers maintaining discipline, but exaggerating in this direction might nevertheless raise difficulties in the future.

At the higher levels, every officer must have a policy, whether they are a commander or a staff officer. They will carry out their policy wisely, loyally, maybe even flexibly—but they will maintain it at all costs. An officer without a policy is not a true officer. He will not have anything to contribute, express, represent, or fight for—and the moment they have nothing to fight for, if conflict is not part of their makeup, they will no longer be a true officer.

I agree that war is a terrible thing and that every human being should recoil from it, but a war of ideas, opinions, and concepts is an enjoyable form of war. At times, it may also involve certain dangers, especially for the individual expressing their opinion. Nevertheless, an officer that tends to avoid ideological conflicts cannot fulfill their duties honorably.

Such a conflict might be dangerous, but the lack of it is many times more so. A struggle might endanger the individual—the staff officer— but the lack of a struggle endangers the entire army. It is essential for the staff officer to "place their policy on the table," and not to keep it to themselves or express it only in private. I am not interested in a policy that takes the form of empty accusations and criticism around the dinner table. A responsible staff officer will state their opinions openly in a cordial, courageous manner at the proper time and place. If they intend to carry out their policies individually and secretly, they are sinning against the army framework. The staff officer's policy must [be] integrated through open, frank discussion with the final policy as determined by the commander. Only then will it have value and make a real contribution.

The Boundaries of the Struggle for Opinions

To what point should the staff officer continue defending their opinions? It is generally acceptable that the struggle should end when the commander arrives at a decision; from that point on, the entire staff [is] expected to act as one body united by one purpose, which is to achieve the objective set by the commander. I would like to express my reservations regarding the above assertion. In principle, it appears solid, but should not necessarily [be] accepted as unconditional.

I will explain this by means of situation assessment. The mission assigned to the staff officer is to express their opinion and offer recommendations to the commander so that they can reach the best decision in given conditions and achieve victory according to their own and the higher level's intentions. Thus the staff officer's role (besides gathering the greatest amount of exact data possible) would be "to continue to defend my opinions even after the decision has been made, so long as it does not interfere with completing the mission as defined by the commander." The great privilege and responsibility that [is] bestowed on the staff officer here is to decide for themselves when their stubbornness is liable to interfere with the completion of the mission and when it is time to ease the pressure.

The commander's formal decision cannot always put an end to the ideological struggle. In most cases, this is what actually happens, especially regarding decisions made on the battlefield, when there is a brief time lapse between decision and implementation. On the other hand, when the decision made at the general staff level and its significance is mainly organizational and long-term, or when the decision is operational but enough time passes before putting it into practice, formal decisions do not necessarily put a stop to discussion. It should definitely continue in a situation where doubts arise regarding an order's suitability to the higher level's intentions, not to mention cases when its legality is in question. In such cases, it is mandatory not to carry out orders, but to act against them.

I have attempted to explain why the staff officer must carefully examine their position and why I cannot accept the convention that the war of ideas within the staff should end when a decision [is] reached. It follows that a staff officer that does not fight for their ideas is not doing their job or supporting their commander, who will not only find the staff officer useless, but also obstructive.

In the modern army, which employs complex, varied weapon systems, the value of staff consultation is on the rise and it is becoming more and more difficult for one person to become expert in all the resources and possibilities at their disposal. One thing is certain—in most cases, the decision reached through intellectual struggle will be preferable to a unilateral decision. The staff officer is part of an organism of opinions and views. It is not their job to make the commander's life easier on the way toward reaching a solution or a decision. It must [be] remembered that the decision must be the optimal one. Decisions sometimes have important, even amazing, implications, but they are also liable to result in disaster. It is the exclusive prerogative and the great privilege of the commander to decide, but this [is] not given lightly or without cost. The staff officer must not be too "good-hearted" and give in to their commander, but must rather continue raising analytical questions and ultimately suggest the best possible solution to the problem under discussion. The staff officer must not take the objective "high ground," later writing in their memoirs: "Today I said such and such and nobody listened . . . I did my part . . . let history be the judge." History will judge such officers harshly and will only pay homage to those who stood and fought. Regarding commanders' morality there is a saying: "You abandoned me today, I will abandon you tomorrow . . . and the next day." Values, the officer class, and the greater good cannot [be] divided.

Knowledge and Opinion

Finally, I would like to comment on the importance of the staff officer's professional knowledge and the connection between knowledge and opinion. Whole sectors of the army become professionalize[d], thus minimizing the significance of opinion that [is] not supported by knowledge. Even if we assume that the art of command [is] grounded more on intuition and understanding than on in-depth knowledge of details, the latter cannot be discounted. Currently, there is a tendency in the staff to seek new ways of exploiting resources and a constant race to overcome limitations and malfunctions that are an integral part of war materials. Furthermore, it is essential to surprise the enemy with innovative procedures and fighting methods. Professional knowledge is paramount for these functions—where each tank is located and when to advance it; with which ammunition to fire. Improvising, while tactically maneuvering large forces and introducing various resources into the battlefield in an unexpected manner, must [be] base[d] on in-depth knowledge that considers the smallest details.

The staff officer who wants their suggestions to [be] taken seriously must therefore ensure that they are well-based and feasible. This will only [be] made possible if the officer devotes considerable time to profound study and analysis of their special area of expertise regarding every new mission.

Summary

If our staff officers become true professionals who know how to formulate opinions, present them, and fight for them, there is a good chance that we will have smoothly functioning, efficient staffs that are capable of absorbing innovations over time. A commander can only be truly successful if they are supported by an excellent staff. A combination of the two will yield a dynamic army, not one that stagnates and "rests on its laurels."

The main challenges that I presented above were how the staff officer should fight for their opinions once they [are] formulated, how they should present them, and up to what point they should continue to fight for them. This is a dilemma that every staff officer and commander at all levels must face every hour of every day. If the guidelines for behavior that I have presented above [are] translated into character traits, they will create a profile of what we require our staff officers to be.

LTG Motta Gur joined the Palmach (the underground-armed group of the Jews in the British Mandate of Palestine). He continued serving in the military during the Israeli War of Independence in 1948.

Gur served in the Paratroopers Brigade most of his career and became one of the symbols of the "red beret" brigade. During the 1950s, he was a company commander under the command of Ariel Sharon. He was wounded during a counter-terror raid in Khan Yunis in 1955 and received a recommendation of honor from Chief of Staff Moshe Dayan. Gur went to study at the Ecole Suerieure de Guerre in Paris.

After two years in France, he returned and was appointed as the commander of the Golani Brigade (1961–1963). In 1965, he was appointed as the head of the operations branch in the general staff of the I.D.F. He later also served as a commander of the I.D.F. commanders' school.

In 1966, Gur was appointed as the commander of the 55th Paratroopers Brigade (Reserve), which he led during the Six-Day War. After the war, he was promoted to Brig. General's rank and was appointed as the I.D.F. commander in the Gaza Strip and northern Sinai Peninsula. In 1969, he was promoted to Maj. General and was appointed as the commander of the northern front. From August 1972 to December 1973, he served as the I.D.F. military attaché at Israel's Washington D.C. embassy. In January 1974, he was reappointed as the commander of the northern front. He was appointed in April 1974 as the 10th I.D.F. Chief of Staff. He served in that position until 1978.

In 1981, he was elected to the Knesset as a member of the Labor Party. Re-elected in 1984, he served as Minister of Health. In April 1988, he was appointed Minister without Portfolio, a position he retained following the 1988 elections until March 1990.

After the Labor Party won the 1922 elections, Gur was appointed Deputy Minister of Defense.

5

The Challenges of Educating and Training I.D.F Officers

Lt. General (Ret.) Moshe Ya'alon

For four years, The I.D.F. has found itself in constant harsh conflict with Palestinian terrorists, who are partially relying on existing infrastructures of the Palestinian Authority and becoming stronger due to no action taken to curb them. In addition, they enjoy the aid, support, and encouragement of certain elements in Syria, Lebanon, and Iran.

The I.D.F.'s current struggle with the Palestinians is very different from former clashes in both character and form. This is not a symmetrical conflict between two armies whose outcome is clear and chiefly based on military achievements, but a "limited struggle" in which the army is part (and not necessarily the main part) of a nationwide effort to achieve a positive outcome.

This new type of conflict also presents new challenges, including the need to cope with fighting in urban areas teeming with civilians, in which terrorist organizations exploit the Palestinian population as "live shields." Thus the military conflict against the Palestinians becomes one where there is the daily need to cope with complex situations of activating forces while preserving morality, maintaining "purity of arms," avoiding the harming of innocent bystanders, and guarding human dignity in a manner requiring discretion and a careful balance between opposing interests and conflicting values.

Thus, a situation has been created in which combat doctrines and methods, war materials, and even organization structure make it impossible to exert the I.D.F.'s full power against its targets and might even render it totally irrelevant. These new challenges demand a constant process of clarifying and understanding reality in order to identify and prepare a response rapid enough to be relevant to military activity—first and foremost through the ongoing education of I.D.F. officers. These officers constitute the army's professional and ethical backbone; they both lead their soldiers and units into battle and are responsible for their professional and ethical training in face of a wide range of operational and moral challenges facing the I.D.F.

I.D.F. Officers

Primarily, we expect I.D.F. officers to be leaders. This was the central characteristic of the officer class even before the I.D.F. came into existence—in the days of the underground armies and the Palmach (Haganah assault companies) and in a period when the Jewish settlement's military experience was still very limited and war materials were scarce. The value of leadership, then as now, proved the crucial element bringing victory in war, conflict, and battle.

It is important to stress that leadership not [be] bestowed upon appointment to a command post, awarding a rank, or completing a course. Leadership [is] acquired, constructed, and formulated over time. Whoever strives to be a commander and a leader who is worthy of their soldier's trust must possess three basic qualities: roots and values, deep thinking, and professionalism.

Values are the foundation upon which doctrine rests. The commander's value system comes to expression in their regarding military service as a calling that encompasses three basic values: defending the state, its civilians, and residents; love of country and loyalty to it; and preserving basic human dignity. Additional values, including responsibility and comradeship, could add adherence to the mission and striving for victory, personal example, preserving the sanctity of human life and preserving purity of weapons, honesty, reliability, discipline, and more.

Only a commander having a broad, steady moral base and with complete understanding of these values and their significance will be able to act accordingly and inculcate them in their subordinates as required.

However, it is not enough to possess the required values. The commander must also know how to exercise discretion and give the proper weight to the values, needs, and constraints with which he must cope. Finding the balance among these various elements is unique to every situation, thus there is no "schoolbook solution" or fixed procedure for each problem. It follows that the commander's values must come to expression through their ability to prioritize and correctly balance among values, needs, and constraints while considering all relevant data.

Although values are the basis, they are not the only requirement for the successful I.D.F. officer and commander. By its very nature, the military framework has a tendency toward conservatism and adhering to fixed conceptions. Although conceptions are vital for the entire system, they are generally the result of applying experience to present reality, thus presenting the danger of preparing for a past war, not for a present or a future one.

Since reality changes constantly and at a rapid pace, I.D.F. officers must know how to adjust and utilize the means at their disposal to adapt to such changes. Primarily, the ability to adjust is conceptual: it must [be] ascertained that the conceptions according to which we function are valid and appropriate for present reality. These conceptions must be clear and based on a common language and set of concepts. In addition, the current situation must not limit military thought—we must free ourselves from the weight of former conceptual and organizational thinking and constantly update the key doctrines according to which we build and activate the forces.

For this to be possible, the army must be led by thinking commanders who constantly learn and develop, leading to an atmosphere of free discussion, ideas, and opinions that promotes critical thinking. The ability of commanders and officers to express their opinions without fear or hesitation is a source of power and pride and should be encouraged. In this context, modesty is preferable over charisma, self-confidence, and intellectual autocracy based on rank, which limits a free exchange of ideas and further development.

Therefore, the army must encourage the lower ranks to level criticism and raise doubts regarding the commander's decisions when this [is] called for, but without forgetting the importance of discipline, which is the basis of any military organization. In other words, the commander must know how to emphasize that alongside freedom of thought there will also be discipline. Only a combination of the two will enable the I.D.F. commander to remain relevant and fulfill their missions, vis-à-vis developing challenges.

The third element completing values and thought is professionalism, expressed as control over the knowledge and skills necessary to carry out tasks by practical application and a constant striving for improvement. In this framework, the I.D.F. officer must have in-depth knowledge of the military profession, including war materials and their activation, navigation, tactics, maneuvers, and combat methods. Therefore, a professional officer will be thorough and will strive to learn the fundamentals of the military profession to its foundation and its practical applications.

The Challenges of Officer Training

We have seen that in the I.D.F. the three basic elements—values and roots, creative thought, and professional education—are vital to the success of the military leader. I.D.F. officers are the infrastructure upon which the army's worth and fitness are grounded. They are the backbone of command, as well as the army's professional and administrative leadership. The officers determine behavioral norms and standards of performance for the entire army.

In this context, the army's training procedures constitute an essential means of establishing the necessary infrastructure for instilling knowledge, values, and learning in I.D.F. officers. The I.D.F. differs from most of the world's leading armies in the way it locates and trains its officers. I.D.F. officers do not decide upon a military career and voluntarily sign up for training in military academies, but rather are identified on the army's initiative from the ranks of soldiers doing compulsory service (at times even before their recruitment), who undergo relatively brief training courses. The major disadvantage of this system is that most officers do not continue on to a lengthy army career. However, its clear advantage over others is the large reservoir of compulsory servicemen from which to select worthy candidates according to their proven leadership qualities in actual battle conditions as they acquire practical military experience.

The importance of officer training in the I.D.F. [is] already identified at its foundation. In 1953, the School for Command and Staff Training [was] established, in which staff work and tactics learned alongside military history, geography, and administration, Middle Eastern Studies, and the specific problems facing the state of Israel. In 1962, the National Security College was founded, which added a dimension of strategic studies linking military practice with political reality and the army with the society and country it serves. In 1999, the Tactical Command College was established, enabling officers who had proven their abilities in junior command positions to undergo in-depth military studies for a two-year period together with studies for a first academic degree.

Summary

The major challenge facing the I.D.F. in training its officers is to create the proper blend of values, thinking skills, and professionalism accomplished through concentrated training programs and ongoing experience in the field, in light of up-to-date concepts relevant to both fighting units and the general staff. The major challenge facing the officers themselves is to develop and preserve values, professionalism, and the ability to reason, learn, and be critical. These are all essential for success as professional, ethical commanders who know how to prioritize and find the balance among values, needs, and constraints. I.D.F. officers and their training programs must be capable of coping with the rapid rate of change in today's reality, while demonstrating the ability to adjust, examine, and adapt conceptual thinking to the challenges of present and future reality.

These processes and challenges do not begin and end with concentrated training efforts, but continue throughout military service and at all levels of the I.D.F. officer class. This vital process enables us, as an army serving in a democratic country, to cope with evolving challenges—even in today's complex reality—and continue to provide optimal security for the nation and its citizens.

Lt. General Moshe Ya'alon recruited to the I.D.F. in 1968 to the paratroopers. He completed his national service in 1971 and retired. He fought in the Yom Kippur War as reservist in the 55th Brigade. After the war, he returned to the I.D.F. as a regular officer in the paratroopers and the special forces. In 1990, Ya'alon was appointed commander of the Paratroopers Brigade, and two years later, became commander of the Judea and Samaria Division. In 1993, he was appointed commander of an I.D.F. training base and commander of an armored division. He commanded the G.H.Q. Special force unit and the paratrooper's brigade. In 1995, Ya'alon was promoted to Maj. General and appointed head of Military Intelligence. In 1998, he was appointed as commanding officer of Israel's Central Command. He was serving in this position when the Second Intifada launched in September 2000.

Ya'alon was appointed Chief of Staff of the Israel Defense Forces (I.D.F.) on 9 July 2002 and served in that position until 1 June 2005.

On 1 June 2005, Ya'alon retired from the army.

In 2013, he was appointed as the Minister for Defense and served in that position until 2016.

6 Directed Command and Mission Command

Colonel (Ret.) Yizhak Ronen

On the battlefield mission, command must have pride of place. At the basis of this idea is the understanding that the battlefield is a realm of uncertainty, so it will be difficult for commander[s] to control what is happening with their subordinates. Thus, it is necessary to adopt an approach by which subordinates accept a mission with all its limitations and take the necessary steps to accomplish it. However, in an era when the higher ranks can track every detail of the lower ranks' activities—at times even better than those in the field—it is tempting to intervene when danger approaches. This means that the supervisory capabilities of available control and supervision systems are liable to lead to headquarters' increasing intervention in a unit's activities.

This phenomenon is liable to dampen junior commanders' innovative spirit causing them not to overly rely on their superiors.

This is what happens when the means (computerized systems) become an end at the expense of efficiently activating the force. In such a situation, any malfunction of the control system can greatly affect performance, to the point where lengthy holdups might lead to missed operational opportunities. One must not forget that our main goal is to reach a decisive outcome, demanding that a force achieve maximal gains at minimal cost and as rapidly as possible. On no account must means [be] turned into ends; information networks are no more than support systems. The problem is that in some areas there is such an obsessive interest in these systems, their performance, and the huge amounts of data that they produce that time is expended that would better be devoted to thought and discussion. Something similar has occurred because of using PowerPoint: too often, it seems that excessive time [is] invested in designing the presentation and too little in discussion and thought regarding the topics raised in it.

In the Six-Day War, when the 38th Division commanded by Arik Sharon was in full swing toward Umm-Qatef, the Southern Command commander, Yeshayahu Gavish approached Sharon and presented him with a dilemma: "Commander-in-Chief Rabin understands from intelligence that it is a heavily fortified compound, so it might be worthwhile waiting another night to allow the air force to soften it up, and only then to proceed." Sharon decided without hesitation that at that stage, when the force was on the move and momentum was at its height, not to delay, whatever the consequences.

What can [be] learned from this? Means are not an end, no matter how much they can offer. Ground battle demands harnessing commanders and soldiers to gaining physical and psychological momentum. Slowing down the force and waiting for the support of means, however effective, does not necessarily work to the good and at times, it is preferable to do without them and rely on the field commander and their soldier.

Sharon was capable of deciding to continue with the 38th Division's momentum, while Gavish was acting according to his lights, not according to mission command. If Gavish had had a clearer picture of what was happening at the compound, he might still have given the order to delay Sharon's forces and bring in the air force. In the end, he decided not to do so because he understood that Sharon had a better understanding of what was happening on the ground, his men's fighting morale, and the best way to go forward.

Figure 5: Gen. Arik Sharon in 1973; I needed to be there in person.

The minute the higher ranks have a clearer picture than their subordinates, there is the danger that mission command will become detailed command and initiative will become the province of headquarters, not commanders. Thus, our era of command and supervision presents a special challenge to preserving the principles of mission command.

Col. Yizhak Ronen served in the I.D.F. Armored Corps in all command posts from Platoon leader to Brigade commander. He was instructor in the Command and Staff College and Chief of Staff of the Northern Command.

His last assignment was the Division Chief of Staff. He was Research Associate at the Begin Heritage Center. Ronen holds a B.A. (distinguished) from the College of Management Academic Studies and an MA from Haifa University.

7 Hypocrisy
Mission Command in the Age of the Strategic Corporal
Lieutenant Yosef Gensburg

One of the main reasons that the I.D.F.'s higher echelons are reluctant to grant responsibility to the lower ranks is their almost total immersion in limited conflicts. In such conflicts, which mainly take place among dense civilian populations and under the watchful eye of the Israeli and foreign press, there might be far-reaching international influence, even on the actions of the simple foot soldier. This has been labeled "the strategic corporal" phenomenon.

Thus through all the years of limited conflicts with the Palestinians, junior commanders have become accustomed to receiving exact instructions from above and have had limited opportunities to show initiative. When those same commanders found themselves in the midst of a real war—in the summer of 2006 in Lebanon—they were suddenly required to show initiative and found it difficult to do so. Undoubtedly, in wartime, there are great advantages to mission command, but if commanders [are] expected to act according to it—to act independently in light of the objective and the senior commander's intentions—they must [be] given the necessary freedom to do so. A commander cannot be expected to suddenly become independent and exercise their initiative after becoming accustomed to being under close supervision, avoiding risks, and not daring to deviate in any way from the instructions handed down to them.

What was stated above presents a knotty dilemma: how can commanders be prepared to act under two diametrically opposed

1   Reprinted with permission from *Ma'arachot*, Vol. 429, 2010.

command styles: the one demanded by routine security measures and the other by conventional warfare.

The Obstacle of Long-Distance Command

Current technology enables senior commanders to watch in real time what is happening on the battlefield without emerging from their rear headquarters. As a result, they are often tempted to intervene in junior commanders' decisions and direct their every move. The Second Lebanon War gave a bad reputation to such a command style; those who adopted it were wryly named "plasma commanders," since they directed fighting from rear headquarters while watching the war on plasma screens.

This command style is problematic. The commander sitting at a distance does indeed have a better view of the battle as a whole, but better decisions [are] often made according to the limited view of the junior commander in the field. It is obvious that command at a distance breaks down hierarchy and cancels out the mission command concept (that supposedly deeply engrained in the I.D.F.). Furthermore, it badly affects junior commanders and is liable to shake their self-confidence. Above all, this is a command method that might be ineffective, as stated by Maj. General Ben-Gal:

> It is difficult to determine military ground operations by remote control. The commander must be present on the field, know where and when to create centers of gravity by amassing forces and when to divide and reorganize them. As was clearly demonstrated in the Second Lebanon War, the miracles of technology create the illusion of being there. Command becomes "virtual" and results in unsatisfactory battle management.

The Goal is to Train Toward Mission Command

The subject of the commander's location on the battlefield has preoccupied the I.D.F. since the Second Lebanon War, so today there is a tendency to return commanders—especially on the brigade level—to the field. Current technology enables them to do this without abandoning the ill-reputed plasma screen, which can fulfill a useful function. Returning the commander to the field is a vital, but not sole, condition for establishing mission command in the I.D.F., since this is more than just an operative technique, but rather an in-depth educational infrastructure, thus demanding proper command training from the lowest levels.

Military professionalism is measured not only by a unit's performance, but by how the whole system functions. At times, forces are required to fight together without having ever met before. This emphasizes the importance of a shared professional lexicon, without which the battlefield becomes a kind of "Tower of Babel" in which forces cannot work together and their effectiveness suffers considerably.

Lieutenant Yosef Gensburg was recruited to the I.D.F. in 2011. He serves as an instructor in the I.D.F. Officers School.

8

The Commanders' Independence

The Future of Mission Command in the Ground Forces[1]

Brig. General Oded Basyuk

Introduction

The introduction of C2 systems in the framework of the "Digital Ground Forces" Project was meant to enable and even encourage activities based on existent leadership principles (i.e., to enable and encourage mission control). The claim examined in the present article is that in current organizational culture, especially in light of commanders' personalities, training, and professional and operational experience, the integration of these systems is liable to increase the likelihood of detailed ground command and a decrease in mission command. Thus, it is a development worthy of consideration, leading to potential changes in commanders' training and command and control doctrine in light of today's reality. Ignoring these matters and not formulating the necessary command and control processes in the era of digital ground forces will make it difficult for commanders to find the proper balance between mission command and detailed command in face of today's rapidly changing battle conditions.

Digital ground forces systems enable all levels to receive information simultaneously, thus allowing the higher ranks to respond immediately to what is happening in the field without waiting for subordinates' reports, as was true in the past. This situation is liable to allow the higher ranks to control the lower ones, thus tipping the balance that makes it possible

1   Reprinted with permission from *Ma'arachot*, Vol. 440, 2011, pp. 28—35.

to maintain the I.D.F.'s preferred command and control doctrine, namely, mission command.[2]

Mission command and detailed command are located at two opposite ends of the command style continuum. The balance between them is liable to shift due to various factors, including the commander's personality or the nature of the mission. It often happens that the same commander in the same exercise or operation acts according to a mixture of command styles; at times, he will lean more toward mission command and at others toward detailed command.

Furthermore, it is difficult in retrospect to determine where to locate the command of a particular mission on the continuum between the two styles, thus different researchers will probably reach different conclusions regarding this matter. It is more fruitful to identify a tendency toward one side or the other of that spectrum, without being overly specific.

2   The article is based, among other things, on the following sources: Brigadier Melvin, "Modern Warfare - Mission Command", *Army Review* 130, 2002, pp. 4-9; Charles C. Krulak, "The Strategic Corporal: Leadership in the Three Block War", *Marine Corps Gazette* 83, 1999, pp. 22-18; David S. Alberts & Richard E. Hayes, *Power to the Edge - Command and Control in the Information Age*, 2003; Eitan Shamir, *Transforming Command*, Stanford University Press, California, 2011; Martin Van Creveld, *The Training of Officers: From Military Professionalism to Irrelevance*, Free Press, New York, 1990; *Dictionary of Military Terms*, 1998 [Hebrew]; Gideon Aqavia, "Mechanized C2 in the Field Forces: Fundamental Problems," *Ma'arachot* 407, June 2006, pp. 17-27 [Hebrew]; Dov Glazer, "On Decentralized Mission Control: Theory and Doctrine," *National Security*, 7 August, 2009, pp. 29-49 [Hebrew]; Doron Rubin, "Mission Command: Its Place and Importance," *Ma'arachot* 318, February 1990, pp. 6-9 [Hebrew]; Ground Forces, *The Concept of Control and Supervision on the Ground*, 2010 [Hebrew]; Ground Forces, Doctrinal Dept., *Ground Forces Operations*, Volume I, Introduction, 2010 [Hebrew]; Hanan Shay (Schwartz), *Commander Training on the Operational Level in the IDF in Light of the Theoretical Fundamentals of the Modern Command Method of Decentralized Mission Command*, IDF Operational Branch, Historical Dept., 1998 [Hebrew]; Meir Finkel, "Striving toward Certainty on the Battlefield and Inherent Dangers," *Ma'arachot* 403-404, December 2005, pp. 4-7 [Hebrew]; The *Influence of the Information Revolution and Changes in the Nature of War on the Command and Control Model in the Air Force*, The College of National Security, 2008 [Hebrew]; Uzi Ben Shalom and Eitan Shamir, "Mission Command: Between Theory and Practice, *Ma'arachot* 418, April 2008, pp. 16-23 [Hebrew]; BG N. "Influence of the Information Revolution on the Command and Control Model in the Air Force," *Ma'arachot* 425, June 2009, pp. 4-11 [Hebrew].

Figure 6: Reprinted with permission from *Ma'arachot*.

The present article focuses on the influence of digital ground forces'
C2 systems on battles commanded on the division, brigade, and battalion
levels during the 2008 Gaza War (Operation "Cast Lead") and two division-
level exercises that were performed using digital ground forces systems. In
the case of the exercises, the analysis will focus on brigades and divisions
and in the Gaza War; it will focus on brigades and battalions. It [is] based
on relevant summaries and more than twenty in-depth interviews with
senior staff officers ranging from battalion commanders to generals.

Detailed Command versus Mission Command

In detailed command, the commander receives direct and ongoing
directions from their superior officers. On the one hand, this command
style affords superiors a large degree of ongoing supervision, but on
the other, it allows the subordinate commander a very small amount of
freedom of action.

In mission command, the higher-level commander determines the objective (what is to be accomplished) and the rationale (why it should be done), as well as the conditions and constraints under which the forces will operate. However, the higher-level commander does not interfere in how the mission is accomplished. This remains the responsibility of the subordinate commander to which the mission has assigned. Mission command obliges junior commanders to understand their missions in depth and adhere to their objectives; even under conditions of pressure and uncertainty, they will strive for contact—at times without receiving authorization—and initiate additional missions as the battle develops. It is clear from the above that mission command demands a high degree of trust among the various levels.

Examining the Parameters

In interviews for the present study, commanders [were] requested to state their position regarding various aspects of command, including where they should be located during the battle. The relationship between the commander and the forward headquarters; the degree to which the commander should intervene in junior commanders' decisions or area of responsibility; the degree of initiative displayed during the battle and the exercises; and to what extent supervision was imposed in order to increase certainty. They also [were] questioned regarding more general matters of language and culture, the influence of C2 systems on doctrinal principles, and the command style implemented during the battles and the exercises. From their total replies, it was possible to determine their position on the scale between mission command and detailed command.

The Commanders' Position

In the division-level maneuvers that [were] analyzed, the brigade commanders replied that the question of where to be located did indeed occupy them. On the one hand, being present in the forward headquarters enabled them to view only partial data. On the other hand, some of the data and output sent to the subordinate command only reached the rear headquarters, where control was more convenient. It may [be] summarized that in the two exercises, the brigade commanders preferred to be in the field. In the subordinate commands, the dilemma was more difficult:

Brig. General: "I cannot say that I did not leave headquarters but I think a division commander has to spend a lot of time in headquarters; control is in headquarters. Leaving headquarters is for [creating] a direct connection and demonstrating leadership."

Colonel (Res.) J., director of the division's combat activities, indicated that the commander's presence in headquarters was no different than it had been in previous maneuvers, but it was more efficient, as gathering information was easier and he could focus on processing and decision-making in light of that information.

Commanders at all levels who participated in the Gaza War generally replied that they spent most of their time with the forces and commanded them directly (with most activity focused on the brigade and its forward command). Some of the brigade commanders moved between the forward command, which was located slightly to the rear, and the front lines. They pointed out that there was a certain advantage in being present at the forward command, especially at phases of the battle when "control was a priority." The commanders indicated that there were relatively few such occasions during the war. Others like Brig. General H. claimed that due to the danger of going against the principle of mission command, they preferred to be with the forces on the battlefield throughout the operation, (i.e. they preferred directly commanding their forces and displaying leadership to being able to control the course of the battle). Battalion commanders could generally [be] found leading their soldiers; the question did not arise of their preferring to spend time at the rear headquarters.

From the commanders' attitude to the question of their location, one cannot observe a rise in patterns of detailed command because of introducing C2 systems (perhaps due to the stringent criticism leveled at commanders after the Second Lebanon War that they had preferred to conduct the fighting from headquarters rather than on the field). In any case, it appeared that most commanders understood that the dilemma of where they should be located had become more acute.

The Relationship between Commander, Forward Command, and Brigade Headquarters

The brigade headquarters' capabilities have improved significantly in several areas:

- Regarding command and control, it has acquired capabilities enabling commanders to view more data.
- Regarding weapon systems, it has acquired the capability to employ a much wider range of resources.
- Regarding intelligence, it has acquired the capability to view more information from various sources and process it effectively. In the future, it [is] expected that a wider range of intelligence methods would be at the disposal of the brigade headquarters, thus further expanding its abilities in that sphere.

The headquarters' ability to direct the battle[3]—especially at the brigade level—has [been] significantly upgraded. Headquarters provide control capabilities only some of which can be realized in the forward command. Due to the additional capabilities that have [been] added, the activities demanded of the operational staff and its head have greatly increased. This might lead to a situation where control overshadows command.

Brig. General Y., a brigade commander in the Gaza Campaign: "The forward command was involved in the fighting. Firing was controlled by

---

3   This refers to the headquarters called "forward," but is generally located at the rear of the forces, as opposed to the forward command, which is generally located at the fighting front.

headquarters . . . the C2 system gave a sense of security and freer firing control."

Maj. General S.: "The process undergone by the headquarters is not only due to C2 . . . in each brigade there is a 'mini firing center' in order to deal with the brigade's sector of responsibility. This is liable to erode mission command, but not necessarily."

Colonel G., a brigade commander of the divisional maneuvers: "The [brigade] headquarters became robust. Thought was given to the position of the Brigade G4 commander or the Intelligence officer."

There has been a shift in the brigade forward command's involvement in the battle due to improvements in C2, but also due to advances in other resources related to C2, such as visual advances in capabilities regarding control [that] make the commander's functioning easier vis-à-vis the fighting force and other headquarters, but also shift the balance between commander and headquarters.

Commanders at all levels indicated that there was ongoing signals communication among commanders and did not generally report that it had lessened. However, in some instances, there was a decrease in communication among command centers, especially on the divisional and brigade levels.

There was also a change in communications between the brigade and battalion headquarters, which was evident mainly in the armored battalions. A major sitting in the command A.P.C. a certain distance from the front assisted the commander in controlling the situation, especially regarding subordinates and neighbors.

Lt. Colonel H., a battalion commander in Operation "Cast Lead": "The battalion is supposedly given firing abilities, but no control systems. These exist in the brigade and the significance of this is clear."

Battalion commanders found it difficult to exploit existing capabilities since they [were]controlled by the level above them. In order to utilize C2 capabilities and effectively communicate with the brigade, they had to seek the assistance of the control A.P.C. and place a senior officer there who possessed the necessary abilities.

Concerning the balance among headquarters, it appeared that there has been a shift in the relationship among the four senior command components on the tactical battlefield: the battalion commander's forward command, the brigade commander, the brigade headquarters, and the division headquarters. The headquarters' C2 capabilities improved, as did the ability to activate weapon systems, thus involving the headquarters more deeply in running the battle. These changes necessarily resulted in an increasing employment of detailed command. Two questions arise: Does the headquarters' current structure reinforce that trend? Is it desirable that it should do so?

The Degree of Involvement

One of the main yardsticks making it possible to determine the type of command—mission command or detailed command—used in a particular formation is the higher echelons' degree of involvement in the activities of the lower ranks. Involvement might be expressed in various ways, not necessarily vis-à-vis a direct subordinate (for example, a brigade commander toward a battalion commander), but by circumventing the middle ranks (for example, if the brigade commander issues orders directly to company commanders) or by means of activating weapon systems in the lower ranks' sphere of responsibility, with or without their consent.

A division commander, Brig. General A.: "The division has the mechanical capability to open fire under the nose of the brigade more successfully than the brigade itself. For complex targets, in the air area of coordination . . . firing from below becomes very ineffective, thus it is not always correct to rely on the lower levels to activate those resources. The ability to attack in the air area of coordination is done by the C2 system . . . A large part of the attacks were carried out in the air area of coordination or very close to it."

Activating various resources by the various headquarters in the air area of coordination where the forces deployed or maneuvering has become a routine occurrence in exercises and battles alike. This is an accepted and widespread practice, especially in asymmetrical scenarios and fighting against irregular enemy forces. The concentration and activation of these capabilities are structurally present at headquarters.

Does the division also intervene in activating firing or other weapon systems in the actual maneuver?

Division commander, Brig. General A.: "We intervened in areas where the brigade did not succeed, at times also in maneuvers. Thus we increased operational effectiveness."

Although there were not many such instances, commanders understood that such interventions might take place according to the forces' capabilities.

Colonel A., Brigade commander in the division exercise: "There was penetration into the brigade's areas of responsibility; for example, if a battalion mounted an attack, firing was performed by combat helicopters . . . battalion commanders are incapable of pinpointing targets with such accuracy. The division attacked based on pinpointing and diagrams presented by the battalion and the brigade . . . There might possibly have been such intervention during maneuvers. The brigade controlled firing in the battalions' sectors."

Did the brigade-level headquarters activate weapon systems in the subordinate units' sectors or intervene in their activities?

Brig. General A., a brigade commander in "Cast Lead": "The C2 system enabled control over small logistic forces…it was easier for me to give the okay to open fire . . . I remember one occasion when a company commander reported that he had entered a minefield. By means of C2 we got him out of there."

Lt. Colonel P., a battalion commander in "Cast Lead": "The brigade headquarters dealt with the surround. Resources activated near our sector. I remember one occasion when our force was hit and a brigade commander intervened directly in the order of battle and the firing systems."

Within the battalion, there was also intervention down to the lowest levels by means of C2. A battalion commander in "Cast Lead," Lt. Colonel A.: "From inside the battalion it was possible to know where the forces were located, giving a sense of security. I intervened relatively frequently in company command—both by direct observation and by C2. At times I also controlled and intervened in the firing of an isolated tank in order to avoid error."

Have the dimensions of intervention grown in the digital ground forces era compared to the pre-digital one? It appears that since the inception of digital ground forces, there have been countless occasions in which the higher ranks activated weapon systems in the lower ranks' areas of responsibility, which would have been impossible without C2 systems or advanced resources of visual. It can thus be concluded that the introduction of C2 encourages detailed command, especially in scenarios where there is asymmetrical fighting.

Degree of Initiative

Both during the exercises and the Gaza War, most of the plans realized without demanding significant changes, thus, there were not many incidents where units on the battalion level and higher took the initiative. (There might of course have been other reasons, but this is not the place to enumerate them.)

A brigade commander in the divisional maneuver, Colonel A.: "The brigade was not challenged by the division to take the initiative. Everything went according to plan."

It is difficult to determine if C2 systems and supervision had a strong impact on initiative. Possibly, the ability to stick to the plan, which generally worked well, removed the need for initiative. The commanders found it difficult to explain why so little initiative was demonstrated, but indicated that examples of it were rare.

The Atmosphere of Security and Supervision, Language and Culture

Colonel (Res.) A., who was responsible for instruction during the divisional exercise: "In the brigades there exists a large professional gap in the ability to utilize resources, including C2 systems. Brigade commanders generally removed from the general picture and the system. They do not exploit the resources provided for them and it is the division that activates them as necessary—at times by direct control . . . from headquarters."

Colonel A., a brigade commander in the divisional exercise: "There was a lot of running around regarding arrows templates . . . The topic of control was very dominant in the division according to people's personalities."

The atmosphere of security and supervision, as well as the culture that has developed in their wake, have been strongly influenced by improvements in C2 capabilities. Many of the commanders interviewed pointed this out, but many others indicated that the culture that developed in the units were strongly influenced by the commander's personality, their behavior in the pre-digital era, and how the battle developed. Many commanders sensed that the level above them—headquarters rather than commanders—gave them a "sense of security." Most of them stressed that this might have [been] endangered by employing mission command:

Colonel A., who took part in the divisional exercise: "There was a feeling in the division that it was aware of the units' state of fitness... in my view this was not always reliable, but it still formed the basis for decision-making."

Brig. General A., the divisional commander: "The illusion of control is very dangerous . . . as a divisional commander I am not interested in where each tank is positioned. I might be more interested in where each company commander is located . . . The developing capability to view everybody . . . creates a situation in which you feel obligated to see everything . . . to deal with everything. This might not yet be possible, but that is the direction we're going in."

Brig. General Y., a brigade commander in "Cast Lead": "The atmosphere of control . . . might result in shirking responsibility. As each commander becomes aware that his superior can view his every action, responsibility will decrease in favor of control."

Not in all the units or in every situation did an atmosphere develop of supervision based on the higher echelons' demand for certainty. Nevertheless, it is definitely possible to indicate that commanders tending toward centralization sensed that they were increasingly able to control matters and impose greater supervision. The findings indicate a trend toward cultural change that is only at its beginning. The risk involved is that certainty might be an illusion and that relying on an integrated picture might be misguided.

The Guiding Principle: Operational Methods Based on Command and Control

Brig. General A., division commander: "The division is more involved in isometric outlines." In battlefields where isometrics are important—in crowded areas in which the enemy is mobile and operates in small bands aided by the civilian population—it is natural that exactitude and control become significant. The use of C2 and visual in "Cast Lead" played an important role in the ability to carry out the operational plan. Among other things, this involved the ability to hit targets that were not visible on the ground, as well as other isometric outlines that practiced nowadays. Command and control systems make it possible to pinpoint and eliminate the enemy while minimizing danger to the forces and lowering the risk of harming uninvolved parties. Control over such resources remains in the hands of the senior ranks. Lt. Colonel P., a brigade commander in "Cast Lead": "As opposed to 'Defensive Shield,' the battalion's sector of responsibility was very limited. Most of it remained on the brigade level and was activated from headquarters by accurate systems."

Will this capability, which has become a reality, increase the control component? Will it reduce the implementation of mission command, as it did in "Cast Lead"? It appears that a number of processes influences this trend:

- The need—especially pressing in isometric conflicts—to act with greater accuracy in order to avoid unnecessary collateral damage and the involvement of overly large forces;
- The capabilities existent today both in the spheres of intelligence and command and control;
- The improved (and supposedly effective) capabilities existent today in the forces' headquarters to carry out these functions.

All of the above enable accurate activation of resources—chiefly from headquarters and under their control—as well as enabling them to take control of activities on various levels when necessary. Thus, for example, it is possible to hit a large number of targets with a relatively low risk to forces. It might soon be possible to construct an entire operation based on such capabilities.

Mission Command versus Detailed Command

In the cases examined for the present study, it is possible to identify processes of increased detailed command alongside the attempt on the part of some commanders to preserve characteristics of mission command. We find ourselves in an interim period in which most commanders did not grow up in the advanced C2 era, but are undergoing a process of absorbing them while still functioning according to their previous experiences, beliefs, and concepts. Nevertheless, the trend is to rely increasingly on these advanced technological possibilities.

The head of the Digital Ground Army Program, Colonel G., related to this topic: "In 'Cast Lead,' C2 penetration was only partial. Not all the forces trained [in it] and conditions were not suitable to exploit fully its capabilities. C2 systems increase the ability to supervise and intervene."

Advanced C2 systems were not ideally suited to that operation and they were not introduced at all levels, but even so, it was obvious that they had an impact on command characteristics. Brig. General Y., a brigade commander in "Cast Lead": "C2 does not replace interpersonal dynamics. Mission command mainly influenced by the commander's personality and C2 can indeed detract from it. In 'Cast Lead,' the division did not intervene and I did not feel that control taken out of my hands. It must be mentioned that there were almost not sector boundaries between adjacent brigades, thus divisional characteristics during the battle were only partial."

The character of the operations and the exercises also made analysis difficult, as they could not [be] defined as total war, but rather as operations in isometric surroundings. Nevertheless, it is possible to draw a number of conclusions regarding future trends.

Lt. Colonel (Res.) A.: "There is a problem with mission command. A brigade commander might act according to mission command, but the [forward] headquarters might cut off from it. The problem is with personnel."

A battalion commander in "Cast Lead": "In the end there was evidence of mission command. The C2 dialogue was with headquarters. I could always put on the brakes vis-à-vis my brigade commander, but I am convinced (and I am not the only one) that if it is possible to take command, then I will do so. The problem lies with many subordinates' lack of professionalism."

Brig. General Y., a brigade commander in "Cast Lead": "C2 supposedly enables mission command. The illusion of control makes it possible to go two ways. C2 might possibly give the security necessary for mission command . . . but when the battle "goes to hell," it might be easy to intervene by means of C2. For control freaks—this is a danger."

Obviously, the commander's personality and that of their subordinates have a great impact, but even commanders who favor mission command—and perhaps chiefly they—understand the potential of detailed command. Lt. Colonel A., an armored battalion commander who fought in "Cast Lead" under the command of an infantry brigade: "C2 is very influential. It is like the introduction of the smart phone. In the past, you would send junior officers to missions and they would carry them out. There was no further communication . . . everything depended on them. Today there are phone calls every five minutes. They do not want to take responsibility and you do not want to risk error. In C2 it is the same, except that in war this does not apply. The command culture in armor is even more supervisory than in infantry."

Do the issues discussed above also apply to previous operations and exercises? Were more principles of mission command realize in them?

In his book,[4] Eitan Shamir claims that principles of decentralized mission command were upheld in "Defensive Shield," but in the Second Lebanon War, they followed to a lesser degree.[5] In those operations, various resources were already in use that enabled headquarters a certain amount of control, but mainly in forward command headquarters and for specific missions only.

The counter-fire controversy is not new and not related only to C2 systems in the framework of the digital ground army. However, although it existed in the past, the discussion of counter-fire versus maneuvering has gained momentum over the years in light of improved technological capabilities and means of control.

4   Eitan Shamir, *Transforming Command*, Stanford University Press, California, 2011.
5   See also Avi Dahan, "Inside Command and Mission Command in the Lebanon War," *Ma'arachot* 409-410, December 2006, pp. 20-25, http://maarachot.idf.il/PDF/FILES/7/112167.pdf [Hebrew]; Moshe Shamir, "The War against Hezbollah: Command and Control Issues," in Zvi Ofer [Ed.] *The Second Lebanon War: Practical Insights*, Ministry of Defense Publishing, 2008 [Hebrew]. Shamir also indicates that in his opinion, there was an improvement in Operation "Cast Lead" in applying mission command, however he did not examine that operation in his study.

Conclusions and Recommendations

Even before the era of the digital ground army, and unrelated to C2 systems past and present, it was difficult to adhere to mission command. There are numerous reasons for this, some of which have [been] presented above: the basic need for control, the fear of taking chances, the army's organizational culture, the need to maintain field security, and the nature of fighting in isometric conflicts, in which every tactical move is liable to be strategically significant. All of these have resulted in only partial realization of mission command. However, it is still considered the preferable command style for three major reasons:

1. The fact that, in any case, commanders have only partial control;
2. The uncertainty of the battlefield;
3. The creative thinking displayed by both sides during the mission command process.

Does the introduction of the digital ground army and improvements in firing and intelligence capabilities form a critical mass that is liable to tip the balance—problematic in any case—between detailed command and mission command?

It is difficult to answer that question unequivocally, but in light of the results presented above, it is possible to assume that we are witnessing an increasing tendency toward detailed command and that there is no guarantee that the balance that existed in the past will continue into the future. For the first time, headquarters can (supposedly) control each tank in the field and the targets at which it aims, view icons and locations of the enemy and of our forces, and the battle scene as viewed by the lower ranks. As things stand today, it must [be] warned that the control component of command is gradually gaining momentum.

We stand at the opening of an era of change and it is essential that we acknowledge this. The characteristics of command and control that already exist today and further developments that [are] anticipated in the nature of war itself might encourage trends of detailed command, which are already evident in the ground forces command. Obviously, the commander's personality has a strong impact on their command style, but this general trend must not [be] ignored. Ignoring it is liable to result in a situation where the two doctrines exist simultaneously, one in the literature and the military academies and the other in the field. Since the commander is likely to act according to command and control doctrine, which has many features of detailed command, they will become accustomed to not taking chances, receiving verification for his actions and making decisions under strict supervision only. However, when they find themselves in a different, more complex, battle situation, they might be lacking in the tools for applying mission command, not having previously created a dialogue with their superiors based on trust.

There will always be some need for mission command, since there will never be full control of the battlefield and total supervisory capabilities. Fighting is influenced by what occurs in people's minds, in other words, it is the result of creative processes, feelings, and instincts,[6] not only the ability to manipulate icons and locations. Technology cannot provide answers for every aspect of war.[7]

At the same time, one cannot ignore the changes taking place in C2 capabilities and the ability to activate numerous resources, including weapons systems; so the question may be asked if the time has not come to examine a new command and control doctrine that integrates the advanced methods existing today. New capabilities in the field of command and control [are] already integrated into mission command[8] and it cannot [be] claimed that C2 only supports other processes and

6   The very concept of the "field forces circle," which makes it sound like a technical process and not a theoretical principle, is problematic.

7   Martin van Creveld, "The Control Crisis," *Ma'arachot* 289-290, October 1983, pp. 12-15, http://maarachot.idf.il/71826-he/Maarachot.aspx [Hebrew].

8   There are many processes that may be made more efficient; for example firing circles, intelligence and procedures of parallel battle management. Adhering to existent doctrine limits the realization of advantages inherent in systems and does not correspond to current trends.

doctrines. The hierarchy is flattening, while networking is influencing the command and control structure.

It is essential that mission command remain the dominant command style on the battlefield. However, it must acquire additional features. As we find ourselves at the height of significant changes, commanders will be obliged to activate more varied and complex methods on the battlefield. Denying this might render us incapable of realizing principles of mission command and will result in a strong inclination toward detailed command. Clearly, existing control capabilities and those that will develop in the future must [be] exploited wisely.

In order to create the necessary blend of these elements, all the components necessary to build a military force must be taken into consideration: doctrine, training and exercises, structure and organization, war materials, and manpower. It is necessary to learn from other armies that combine detailed command and mission command and to create a similar situation for our ground army by adjusting command and control doctrine to accommodate advanced capabilities in the areas of control and activation. An attempt to limit the use of those resources will only be temporary and will not solve the problem. The idea that existing command and control doctrine can be adapted to the digital ground army has not proven itself.[9] Ignoring change will merely widen the gap between the effort to preserve what really occurs on the battlefield and mission command. A new doctrine should [be] considered that would redefine some of these concepts, combine them, and make it possible to integrate features of mission command with those of detailed command. After the new doctrine is defined, relevant processes of training and maneuvers must follow, which today are not always present, although the need for them is increasing unrelated to advances in C2 technology. However, training and maneuvers are

9   Many processes are subject to change and greater efficiency. There is room for structural change in the area of intelligence gathering and fire control in the divisional headquarters. Such change will make it possible to achieve efficiency and flexibility in divisional fire control.

not enough. Additional steps must [be] take[n] in order to enable the retention of principles of mission command.

In addition, systems and capabilities should [be] considered that would narrow the gap between the commander, their forward command, and the headquarters at the rear. The forward commands' capabilities should be strengthened, both technologically and structurally. It should be able to respond quickly by means of a few staff officers that enjoy a high level of security. For example, a small forward command might [be] create[d], made up of two or three well-protected vehicles. In addition, the forces headquarters should reduce in size, rendering it more efficient and flexible and placing it in closer proximity to the commander.

An additional change that is worth considering is creating "controlling" battalions that will assist the commander in controlling the battalion from a rear position in the same way that the chief staff officer assists the brigade commander. The introduction of such a function will help narrow the gaps between the levels. In addition, it is important to create a functional concept that will make it possible to lower tensions between detailed command and control and mission command.

Selecting suitable manpower for headquarters could have a critical influence on the ability to exploit the capabilities of C2 systems to the fullest. Appointing operational officers who are directly involved in directing the battle and activating various resources on the battlefield will demand new training programs and skilled manpower. In this context, it must [be] mentioned that operational staff officers must continue to remain in close contact with the commander.

Military culture has a critical impact: the more we succeed in inculcating the features of mission command in daily routine and training maneuvers, the better we will succeed in maintaining it on the battlefield.

The more we introduce features of detailed command into the army's daily routine and training exercises, the harder it will be to activate mission command on the battlefield. As was the case in the pre-digital era, the commander must impose their authority unrelated to the advantages provided by information systems.

In an aside, it must [be] note[d] that the manner in which we absorb new weapon systems—from recognizing the need for them to actually implementing them—is also liable to widen the gap between their activation and combat doctrines. In an era where technology is developing at a dizzying rate and change is rapid and significant, we must develop a framework that will be capable of including all these components

We are at the dawning of an era of critical changes in the I.D.F. in general and in the ground forces in particular. Even if these developments [are] chiefly based on technology, this does not lessen the role played by people in making decisions on the battlefield. Nevertheless, they have the potential to significantly change the way those decisions are made. Thus, change must carefully [be] planned for and assimilated.

Brig. General Oded Basyuk recruited to the I.D.F. Armored Corps in 1991, commanded the tank battalion and tank brigade, and later was the head of the Operation Department at the I.D.F. J3 Division when he then commanded a Division at the Goal Heights. He holds an LL.B in law and an M.A. in business management.

9

Mission Command and Logistics

Why is it so Difficult?

Lt. Colonel Ariel Amihai

Mission command is a deceptively simple, widespread commanding style that is applied in peacetime and in wartime, as well as in the private sector, in business, or in government institutions. This being so, why have so many articles been written and seminars held to discuss it? What is more trivial than the person in charge saying what must be done and allowing their subordinates to carry it out? Why do more and more logistics officers and commanders employ detailed command? In the I.D.F. Ground Force's doctrinal manual, *Command and Control for Ground Operations* (February 2016, Hebrew), a distinction is made between detailed command (also called "command contingent on permission"—command that depends on the superior's approval) and mission command. Command contingent on permission is based on the assumption that every commander must formulate an exact picture of their subordinates' situation that resembles or is identical to what they report to them. In other words, they must pay attention to every detail when estimating the situation. On the other hand, the mission command method is based on the assumption that every officer or soldier is most suited to their role; thus, it is unnecessary for the officer to report every small step they take.

Since logistics must fulfill material needs that can generally be esti-mated beforehand, the logistic array must gather exact data. This process is the responsibility of logistics commanders on all levels. Furthermore, since logistics is a continual chain, the logistics staff officer of the unit is responsible for providing a logistic answer for all the units subordinate to the level at which they operate and must ensure that the planned solution and that which is provided in reality fulfill all operational demands. The goal of gathering exact logistic information and ensuring supplies to each unit often demands shifting command styles—from mission command to detailed command—or even extending supervision beyond what is necessary (i.e., "command contingent on permission"). In the present chapter, I will attempt to analyze the challenges and difficulties facing us when performing command and control and ways of coping with them.

The Headquarters Company as an Example of the Logistics Officer's Techniques and Abilities

Commanding a headquarters company is not an easy challenge. It is generally the largest one in the battalion; in addition, the manpower serving in it tends to be heterogeneous. Like any other company, it includes soldiers, NCOs, and officers, but the NCOs serving in it are generally older; in some cases, their ages are twice that of their commanders.

Furthermore, while all other companies train together, the training of company headquarters officers is individual, as every functionary is trained according to their specific role. Thus, the headquarters company's commander must devote much attention to new recruits to the company and integrate them into the daily routine by personally accompanying and training them.

The logistics officer is under constant pressure due to their being subordinate to two bodies simultaneously. On the one hand, they are subordinate to the operational branch commander and on the other hand, to the professionals in their field. For example, on the battalion level, the HQ commander is under the jurisdiction of the battalion commander but is professionally responsible and receives commands from the brigade logistics officer. This demands that they maneuver between these two bodies. They must advise the operational branch commander of the logistic repercussions of the operational plan that they have formulated. To their superior on the professional level, the commander of the logistic unit must present the program they formulated to meet the unit's requirements in the best possible way.

The headquarters company is responsible for providing logistic assistance. It is a unit that includes many professionals; each is an expert in their field, as opposed to the operational company in which all participants have the same profession: combat fighter. At the beginning of their army career, the operational company commander was a regular soldier; later, they were a section and platoon commander before they were appointed company commander. It follows from this that they have considerable knowledge and experience in all the operational aspects of commanding a company.

On the other hand, the headquarters company commander cannot possibly encompass all their subordinates' fields of expertise. Thus, they must constantly consult their staff in order to understand the significance of the decisions they must make. They need not be an expert in every field, but they must know how to activate their soldiers, be aware of their strengths and weaknesses, and give them independence in a mission command framework.

Why is it so Difficult to Apply Mission Command and What are the Dangers of Abandoning It?

1. The ability to delegate authority in light of the principle of personal responsibility

Not every position holder is capable of "letting go" and delegating authority, thus losing a measure of control. We often encounter individuals who are "control freaks," who wish everything to be carried out exactly as they wish. In addition, in commanders' courses we are taught to take personal responsibility. If you are in command, it means that you are responsible; thus, many commanders assume that they need to control whatever takes place in their formation. This especially holds true for logistics officers who provide a service, which includes over areas in which things can go very wrong, their responsibility for complex day-to-day solutions (food, transport, fuel, maintenance, etc.) and with a heterogeneous staff.

2. Placing trust in subordinates

In order to activate mission command, one must trust one's subordinates and rely on them to carry out their tasks properly. One must know their qualities, weaknesses, strengths, and abilities to ensure that they will calmly complete their orders, to define their goals, and leave it to them to decide how to achieve them. Quite often, especially in the I.D.F., when we work with subordinate officers for a period of no more than one or two years, we do not create this bond of trust, so we act according to "command contingent on permission."

3. Mission command demands being prepared for subordinates' mistakes

In many cases, the superior officer has considerable experience and much responsibility; thus, they attempt to minimize the chance of mistakes. The superior officer will often act according to detailed command due to a lack of patience with errors and the attitude that they are unacceptable. As a result, a stifling atmosphere is created in which subordinates "tiptoe" around their leader. Furthermore, attempts by the media and senior officers to place blame after any unfortunate event, whether it be routine or an emergency, reduce the readiness to apply mission command (this is in contrast to the familiar "blame the sentry" complex, meaning commanders' escaping blame and transferring it to the lowest echelons).

4. More occupied with limited local operations than with prolonged wars involving several fronts

This situation, together with advanced technological command and control systems, enables senior officers to control the forces' position, observe them in action, and make direct decisions regarding the operation by circumventing a chain of command that limits their ability to make independent decisions.

5. Inefficient exploitation of time

If they do not apply mission command principles, much of the senior officer's time is consumed by consultations, presentations, and authorization, as each matter must be presented to superiors and be approved through a long series of discussions at every level. For example, if a company major prepares a PowerPoint presentation for the head of the General Headquarters, they must have it approved by the Colonel, Major General, or someone higher up. This is an example of how valuable time is wasted by headquarters officers in preparing material for a preliminary discussion rather than fulfilling their true roles: gathering data, describing the situation, formulating rationales for the mission and operational suggestions for the commander, and, finally, providing logistic solutions for the force.

Despite the Difficulties, How Can We Apply Mission Command in Times of Peace and Emergency?

1. Trust in subordinates

All commanders (whether operational or logistic) have the crucial task of creating mutual trust with their subordinates. The commander generally appoints their subordinates based on their character, professional level, abilities, and leadership qualities, as expressed in the evaluations of their former commanders. The commander must begin with the assumption that their soldiers are capable of doing what is demanded of them. In addition, they must create a work atmosphere of trust, openness, and respect that encourages subordinates to believe in their capabilities.

2. Properly defining the mission at hand

The commander must be responsible for presenting their logistics policy (in addition to professional matters) as defined in the I.D.F. doctrinal manual, "The work of commander and headquarters" (Hebrew). This manual determines that "the support statement defines the commander's general policy, as determined by the commander himself or the operations branch officer appointed by him, as well as constraints or aspects that the commander wishes not to leave to the discretion of those receiving the orders."

Many commanders "commit the sin" of leaving it to other logistic officers to write their professional mission statement. This resembles a customer who says to a waiter, "Bring me good, satisfying food" and expects the waiter to know exactly what food will appeal to them. In their statement, the commander defines their expectations of the assisting forces[1] to allow them to carry them out. For example, "The force will need to fight for twenty-four consecutive hours without meeting up with any logistics units. After twenty-four hours, they expected to reach point A, where they will require supplies. You will be assisted in this task by Company A." By means of defining the mission, apportioning the means, and defining the sector's boundaries if necessary, the commander will enable the logistics unit to supply the necessary provisions.

3. Professionalism and a common language

Professionalism, knowledge of military terms, and a common language must be the logistics officer's guiding principles.

---

1   The command and control principle is "the one being assisted dictates to the assistant". This principle states that the goals and activities of the assisting force (in our case, the logistics force) are derive from the needs of the recipient of the assistance (in our case, the operational force).

For example, when the logistics officer reports to the commander that the arrival of the forces in the allotted assembly area will culminate at a specific hour, it is imperative that both officers communicate in a common language. It must be understood if this culmination is meant to include simply arrival at the assembly area, or, alternatively, it may include the unloading of all relevant vehicles or even readiness of all forces for further movement. Without this mutually understood language, it would be necessary to employ "detailed command" in this scenario as well as others like it.

4. Supervision

Exercising mission command does not mean a lack of supervision. However, its function should not be to control every detail of the subordinates' plan but to ensure that it corresponds with the commander's intentions. Supervision must reflect the logistic plan or the individual supervised in order to ensure that there is a common language between commander and the logistics officer. The officer vetted will gladly receive supervision that aims at improving the solution, coordinating components, and offering aid. When our children get something wrong, instead of getting angry, we might say: "If you did not understand, it must be because I did not explain myself properly." If the plan is unsuccessful, the commander must not only blame the logistics officer (thus turning their relationship into one of "command contingent on permission"). Rather, they must analyze what was unclear in the definition of the mission or where the professional gap occurred in order to bridge it in the future.

5. Being prepared for error

How often do we allow ourselves to take chances in order to learn and develop? How many times must we stumble and fall before we can walk? Do we afford such opportunities to our subordinates in order to empower and teach them so that they will develop successfully? Of course, this should preferably take place in more relaxed times, but it will definitely result in high-quality officers who are confident in their abilities.

6. Proper training

Commanders should be fully conversant with the command styles that are currently in use and the advantages and disadvantages of each in order to adapt their style to each situation and personality. In my view, the command style should be studied down to the level of the soldier; furthermore, it should be part of the ongoing dialogue in the unit so that an officer can tell their commander: "For this mission I will act according to detailed command due to its sensitive nature." Even a soldier who is used to receiving detailed command missions should be able, in times of emergency, to do what is required without direction, supervision, or accompaniment by his superior. For instance, the slogan inculcated by logistics corps commanders "to seek contact with our forces"[2]—meaning that we must provide assistance to the fighting forces in dangerous areas— should not only apply to the commanding officer. It should also apply to the oil tanker driver and the tank mechanic. The soldiers taking the greatest risks are not officers situated a few miles to the rear of the battle or sitting in a relatively protected A.P.C., but those accompanying the forces: the driver or the mechanic, alongside officers and NCOs. We must teach those functionaries who are used to receiving detailed commands to understand that in emergencies, they will need to independently make the necessary adjustments, face difficulties, and carry out their tasks at any cost.

---

2   "Seeking contact with our forces" is the logistic equivalent of "seeking contact with the enemy," a fundamental principal guiding the fighting forces. Since the logistic effort directed toward our own forces and not toward the enemy, it must provide all that is necessary regarding ammunition and all other requirements under any conditions.

Conclusion

The present chapter attempted to present today's challenges in preserving mission command as a viable and necessary command system. It is essential to recognize the existing difficulties in applying this method and proper training, together with the other points that are raised above, is essential for command in times of peace and emergency up to the present day.

This method should not be applied in times of emergency only. The command and control principle of "training as we will fight" should be applied even in peacetime. Training personnel, their behavior, and their relationship with their superiors in peacetime will most likely lead to a similar situation in emergencies. A subordinate who finds it necessary to receive verification from their commanding officers for every action in peacetime will probably not be mentally flexible enough to act independently under battle conditions. Obviously, the commander is unable to march alongside their soldiers in order to check that they are on the right track, technically sound, acting as they should, and providing the necessary logistic solutions.

In order for a logistics officer to provide a unit with the best service, the commander must provide clear definitions, allow freedom of action, and supervise the planned and actual logistic mission. It may be assumed that the longer the mutual effort between the provider and recipient of the assistance, the less it will be necessary to adhere to these rules. But, especially in times of pressure in critical missions, not to mention battle conditions, the recipient of the assistance must ensure that they fully explained their demands and that the provider has understood them correctly.

It is possible to apply the principles defined above in any relationship between a commander and their subordinates or between any higher and lower authority. Logistics commanders and officers must be fully cognizant of the plan of action, but they must also leave their subordinates the freedom to carry it out. Despite their desire to offer solutions to every level of the logistic chain, they must remember that the responsibility for logistic solutions at times of warfare falls on the commander at that level, who, together with the logistics officer, must provide the necessary solutions.

Lt. Colonel Ariel Amihai recruited to the I.D.F. in 1992, served as a logistics officer in the battalion and brigade level, then moved to the doctrine department at the Logistics Corps and as the head of Logistics Technology Systems at the Ground Forces Command. Amihai is currently serving as head of the technologists' weaponry section of the I.D.F. Ground Forces Weapons Systems Department.

Amihai holds a B.A. in Logistics and Economy and an M.A. in Business Management.

10

Mission Command on the Tactical Level

Points for its Successful Application[1]

Lt. Colonel Idan Morag

If commanders are informed regarding supervision and reportage, recognize the importance of differences of opinion as a basic condition, and coordinate expectations with their juniors, it will be possible to improve the application of mission command in the I.D.F.

Over the years, several command and control approaches have developed, the two major ones being mission command and detailed command. Since the I.D.F.'s beginnings, the preferred command style has been mission command, apparently due to its roots in the Palmach (Haganah assault companies) and the underground fighting organizations. In 2013, it [was] officially pronounced "the I.D.F.'s preferred command doctrine."

Although many years have passed, Dr. Eitan Shamir wrote in his book on the subject that the I.D.F. (like some foreign armies) has not managed to fully inculcate mission command into its ranks. The book [was] written in [the] wake of an in-depth study and presents reasons for partial or negligible success in establishing this doctrine.

The first question that must be asked is "What is mission command?" According to I.D.F. doctrine, it is a command and control style in which the senior commander determines the purpose of the mission and the constraints under which their juniors must function, without dictating how the mission is to [be] carried out. The ground forces' battle doctrine demands junior commanders "to define the operational conception and method of carrying out the mission and take the initiative in achieving it."

1   Reprinted with permission from *Ma'arachot*, 466-467, 2016.

In officer training courses from the platoon to the battalion level, lessons and discussions [are] conducted about mission command and ways of applying it. In courses at officers' training camp and the Tactical Command College, trainees [are] exposed to the conditions necessary for mission command to exist: an agreed battle doctrine; common thought processes; mutual trust and reliability; junior commanders' readiness to initiate and take responsibility; senior commanders' willingness to accept the possibility of mistakes and failure; and the courage to admit mistakes.

In addition, principles are defined that senior and junior commanders must learn and apply, each at his own level: to clearly define the mission and its objectives; to impose a minimum of limitations, conditions, and restrictions; to place trust in subordinates; and to maintain the chain of command. It is demanded of the junior to adhere to the mission in light of its objectives, take full responsibility and exploit the freedom he has been granted.

Figure 7: Reprinted with permission from *Ma'arachot*.

In the staff and command officers' training course, the "how" is defined and it is clearly stated that the commander's role is to determine the mission. However, the question may [be] asked: "If the I.D.F.'s battle doctrine prioritizes mission command and the army educates toward it, why it is only partially applied?" In his book, Shamir suggests possible answers, starting from shifts in public opinion vis-à-vis the army and its blunders, going on to inadequate professional training, and ending with the introduction of innovative on-line command and control systems. Both Shamir and other senior officers point at the increase in routine security measures performed by commanders as the central reason for an incomplete application of mission command or a total abandonment of it.

In my opinion, the factors mentioned above present a challenge to mission command, but the best chance of applying it does not lie in attempting to overcome these problems. Firstly, from my experiences as a commander and conversations with colonel colleagues and other commanders, I have learned that in most routine security measures, commanders today have a large degree of freedom. In my view, preferring detailed command in routine security measures is unrelated to conditions on the ground. Secondly, online command and control systems are axiomatic on the modern battlefield chiefly due to their tremendous operational advantages. These innovations will continue to serve the I.D.F. for many years to come and we must find ways of overcoming the challenges they present. I think that the main gap in applying mission command stems from mistaken conceptualization and unfocused learning. Undoubtedly, a professional standard and common battle doctrine combined with mutual trust and reliability are basic conditions for its existence. At the same time, I am convinced that we must more clearly explain to commanders and their subordinates what [is] expected of them. The major question that [is] raised when discussing mission command is whether the commander is prepared to accept the possibility of error. This is a correct and necessary point, but generally, it is not realize[d] in full. Accepting error [is] mainly linked with performance and battle management. In other words, application centers on the question of how we react to commanders' blunders and failure to carry out missions.

However, what about the earlier stage of planning and battle management? At that stage, most criticism [is] directed at not applying mission command. The answer at the planning stage is not to accept mistakes, but to accept differences. The commander should be willing to accept and authorize a different plan from the one they formulated themselves. They will generally assume that the plan they propose is preferable to others. According to I.D.F. doctrine, mission command stipulates that the senior commander will determine the mission and the limitations within which subordinates [are] meant to carry it out, emphasize ways of successful application, examine that the plan is professionally sound, and approve it, even though they estimate that the outcomes will be different from those of their own plan.

Subordinates must [be] taught that the senior commander's criticisms do not necessarily fall under the category of detailed command, but are just as relevant to mission command. Another element that is lacking in today's educational system is supervision and reportage. The supervision process [is] built into command. The constant need for certainty strengthens the commander's need to know what is happening under them, the result being constant tracking of the junior to the point where they feel persecuted. Mission command does not mean anarchy; the commander is responsible for controlling events, thus supervision is vital. Commanders and their juniors must [be] taught how to cope with the tension between a correct measure of control and brutal intervention in juniors' activities. Commanders must be committed to two combined processes: coping with uncertain conditions and reasonable supervision.

Such behavior is also appropriate to other situations, including training, routine security measures, or any other type of mission. Subordinates must also maintain two combined processes: accepting supervision as part of mission command and not as a sign of detailed command, while being responsible for keeping their commander up to date by ongoing reportage. This will strengthen the commander's faith in them and will prevent a sense of lack of control that will increase supervision to an unacceptable level.

In conclusion, mission command has been determined to be the I.D.F.'s preferred command style. It [is] taught by officers at all levels on the understanding that it will [be] applied in most types of operations. However, the manner and frequency of its application is currently inadequate. I am convinced that if commanders [are] made aware of important aspects, including ways of supervision and reportage and the importance of accepting differences as a basic requirement, and if there is coordination of expectations between senior and junior commanders, it will be possible to realize more fully mission command in the I.D.F.

Lt. Colonel Idan Morag recruited to the I.D.F. Armored Corps in 2000 and commanded the tank company and tank battalion. As Company Commander during the Second War in Lebanon (2006), he was awarded the Medal of Distinguished Service.

Part 4

Mission Command Put to Test

In this section, commanders describe their battle experience, when they faced the "real world" when they have a mission, when they lead their soldiers, and when there is a battle to win—and winning is their responsibility and no one else's.

1

First Missile Boat Battle

The Israeli Navy vs. the Syrian Navy, 1973

Captain (Ret.) Ehud Erell (Navy)

The Battle of Latakia was the first battle fought on both sides with ship-to-ship missiles. Until that time, it was mainly guns. There were cases where the other side had some form of cruise missile that did work; unfortunately, that also happened to us. However, this was the first time in naval history that a battle was fought with missiles on both sides, and it was a completely different ball game. This small point on earth was one of the focal points between East and West during the Cold War, and the Soviets were arming Syria and especially Egypt. On the naval side, which was very tough for us, they were arming them with post-World War II destroyers. At that time, they were called "Skoryy" destroyers. The Skoryy were a naval threat and we did not have anything to combat them. What we tried to do was develop something that could reach further than their guns. Their gun range was something around 14–16 kilometers, and we did not have anything with which to hit those destroyers. Therefore, the navy said, "Let us try and develop it." Moreover, the development that went on the Gabriel—as we called the missile that was developed— required a platform that could carry it. What the Russians did was take a torpedo boat and mount two Styx missiles on it; there was radar, which was also used by torpedo boats, but I am not even sure if they could detect any targets at the missile's maximum range. The idea was to equip it with Gabriel missiles to reinforce the remote-controlled 40-millimeter short-range anti-aircraft guns with effective range around 4,000 meters against surface targets. The ships were called *Saar 2* (Saar in Hebrew means "tempest"). They were of superior speed and could do up to 40 knots; that particular model could do around 30 knots. They had a very good range. It could remain at 30 knots for 30 hours; in other words, it could travel up to 900 miles. The ranges to our enemy ports are up to 250 miles, so 900 miles meant that we could reach Italy.

They had very good sea keeping and excellent power. The main thing to emphasize was their state-of-the-art surveillance system and electronic countermeasures; everything that the best minds could develop was put into those vessels. It was a completely new concept. At the time the Gabriel was developed, its main target was the destroyer. Then, on one of the firings tests, we were using it to shoot at one of our old destroyers that was earmarked for sinking. The wind changed the aspect of the target (nobody was on the ship, of course) and the forward part was very narrow and had a flare. There was a debate whether to launch the missile because it was not intended for such a narrow target. It was decided to launch it anyway, and the missile hit right where the anchor comes out. The engineers began checking if this was a freak occurrence, and it turned out that the seeker worked best if the target was smaller, so there was no problem trying to decide where to hit because here, it was very clear.

Therefore, we understood that we had an effective weapon against small boats. We checked it on an old torpedo boat that was being operated at that time by a crew member who put it forward to 30 knots, jumped overboard, and ran. When he was clear, the missile was fired, and only some pieces of wood on the water's surface remained of that old torpedo boat.

So we knew that we had a very good weapon, but it had a range of 20 kilometers only. The Russians then went a little further. They took another torpedo boat, but a better one, that was almost the size of our vessel but that could carry four Styx missiles. They also had a remote-controlled 30-millimeter gun and fire control radar. This fight was starting to be serious. The focus, little by little, showed that this was our target, not the destroyer.

That is the point to which we had arrived on the day before the '73 war broke out.

The *Saar 4* was an excellent, beautiful boat, the best I ever commanded. It had 76-millimeter guns fore and aft, and it could carry eight of those missiles. It was a very strong vessel that could sail the Red Sea at 30 knots (a distance of 1,000 miles), remain there, and then return. It was a great boat.

As for the operational concept, the Styx has a range of 50 kilometers, and the Gabriel, a range of only 20 kilometers. We had to do something about that. The idea was to use our superiority in detection, to be able to detect the enemy without their detecting us by using our radar detection systems, call the Air Force to help us cross the gap, and then close within range. If the Air Force did their job, great; if not, we would finish them off. That was the basic idea; our idea was that they would see the chaff before they saw us. It worked beautifully.

It carried on the wind and spread around. Nevertheless, we had rockets that could travel about 12 kilometers, very far from us. The idea was to use tactical chaff to make them expend their missiles. We did everything we could to develop that capability as well as did a lot of training. We knew that the Styx flew and searched for its target from infinity towards itself. Our concept was to open the chaff very close to the vessel; if it was already locked on us, we would rush towards the enemy missile at 40 knots and the chaff would stay behind so the missile would remain with that. It worked, except when you did not fire it in the right direction with the wind. You had to know where the wind was coming from, because if it was coming from the direction you were firing, it would land here, as indeed happened during the war. This demanded tight control. From all our simulations and hours of training, we understood that it was going to be very difficult to understand what was what.

Figure 8: *Mivtach* anti-submarine missile boat

The war broke on October 6, 1973. It was a Saturday and Yom Kippur, the Day of Atonement, which is the holiest day of the year on the Jewish calendar. At 2 o'clock in the afternoon, both the Syrians and the Egyptians mounted a major attack on Israel's borders. It was a surprise attack; everyone was surprised, apart from the navy. For seven weeks, the chief naval intelligence officer kept saying "Listen, there's going to be a war; they're preparing for war." He had his own signs. For instance, the Russians had left Port Said. He said that this was a sure sign of war. He observed other signs but could not convince the rest of the intelligence community that war was brewing.

He approached the commander in chief of the navy, who went with him to the commander in chief of the armed forces and demanded that he listen to him. However, the armed forces' commander in chief did not accept his word either because he had his own intelligence officers.

Finally, the commander in chief of the armed forces told the naval commander in chief, "If you think there's going to be war, prepare for it." Just like that.

I mention this because two days earlier, the navy had performed a major drill that had nothing to do with the war.

Two of the boats, Saar 4's that were meant to go around Africa, were on a training voyage in the Mediterranean. On the way back, they played the role of enemy destroyers attacking Israel, and the whole navy was out. This took place on Thursday night, and the navy out all night. On Friday, we were all at a debriefing, and the commander of the flotilla came in at 7 o'clock that morning and said that we were not going to have a debriefing. We were to go to our boats and prepare them for war. By this, he meant that there was a lot of logistic preparation, as we had to exchange 20 mm guns for missiles. When you install missiles, you have to check them, and it is a long process. We never went out with a full load of fuel or ammunition because it just made the boats heavier. At about midnight between Friday and Saturday, we were ready. At that time, I was the flotilla's operations officer. I went to sleep at about midnight. At 4 o'clock in the morning, I was woken up (I was a mere lieutenant at the time) and told to return to base, so I did. I arrived at the base quickly (I lived nearby), and the word was that today, Saturday, at 6 o'clock in the evening, there was going to be a war.

From what I learned much later, at that time the head of the Mossad made a trip to London to meet with the top spy that we then had, and the spy told the head of the Mossad about the intended attack at 6 o'clock p.m. What we decided to do was send out the boats, some of them towards the Egyptian side, some of them towards Syria, because we didn't know exactly what was going to happen—we didn't even know for sure if there was going to be a war. This was strictly a navy decision, having nothing to do with the general staff. Of course, some boats were left behind to guard against submarines, and the flotilla commander did not want to decide which way he was going, so he ordered that one boat be left in port.

I remember that, at around half past one in the afternoon, I was walking on the pier and it was deserted; there was only that one boat waiting for us. I told the flotilla commander, "Listen, if we don't leave now, we'll miss the war." So we left, and he was trying to decide which way to go; should it be Egypt? Nobody could have known that there was going to be such a long war, so he could not decide. Mentally flipping a coin, he opted for joining the squadron that had been sent in the direction of Syria. The squadron leader was on a Saar 1 boat because it had more room since there were no armaments. That was a big mistake: no missiles, nothing. There were actually three missile boats out there waiting for orders. He decided to join them. By this time, it was half past five.

When we left the port, the sirens sounded and we understood that we were at war. After a short while, it was announced from headquarters that there had been an invasion and we were involved in a total war. On the spot, the commander decided that we were not staying there but were going north. That was not the original plan, which was to wait somewhere on the Lebanese coast. Lebanon did not have a navy to speak of, so we could do what we liked there. However, going up as far as Syria meant that we would be out of the range of air cover. When we were opposite the southern Syrian port of Tartus, we began intercepting Syrian missile boats' radars. The commander was very aggressive. It was 7 o'clock p.m. and already very dark; we were about 40 miles offshore. According to protocol, we immediately called up the air force, and, of course, they did not come and I could not blame them; they had a lot on their hands.

The flotilla commander quickly realized that, and, after about five minutes, he said, "I'm not going to wait for them." In spite of all the rules of engagement and everything, we decided to attack. We started rushing in, and with the tactical staff, at a distance where we should have seen them on the screens, we switched on our radar but saw nothing. We lost heart a little, stopped the attack, and regrouped. It was decided to turn the other way and try to surprise them by arriving from the Turkish side, not from the south, from Israel. At around 10:30 p.m., we did not see the target at first but only got a radar signal of a very small torpedo boat that we knew was a Soviet K-123 boat, as we knew that the Syrians had it. It was a very small vessel, about 16 meters long traveling at a high speed of 50 knots with a small crew. We did not understand what it was doing there, such a small vessel and so far out, so it was very easy. We fired a warning shot, and the person made the big mistake of returning fire. They got every kind of bullet that we could put into them and was stopped in their tracks. We still did not know what was going to happen, but the flotilla commander said, "I'm not leaving this guy afloat."

It was made of wood; it did not sink easily, so one of the boats was dispatched to sink it by machine gun. The rest of us started going a little to the northwest, towards the port of Latakia because we supposed that they had warned off the Syrians, which they had. Now, we were down to three missile boats and a gunboat. Again, that was a mistake, because we should have left the gunboat, as it was something we could easily do without.

Next, there was a new target, immediately identified as a T42, a mine-laying vessel, maybe also a minesweeper, and so it was worth firing a missile. Incidentally, since we were not yet sure how the missiles would perform, we did not shoot that many during training. The normal procedure was to shoot one and shoot a second one a minute later, even before the first one hit its target. In this case, because we knew it was not a missile boat, we decided to waste only one missile on it. It was hit and became dead in the water, but it did not sink. Later, the fleet's sweeper—the boat we had left behind to finish off the torpedo boat—arrived and finished it off. Then, Styx missiles appeared from the southeast. I believe now that the radars that we had intercepted earlier were missile boats that were just leaving port. Maybe our triangulation was not very accurate and they went out on patrol. They did not know anything about us until they heard from the torpedo boat. When we started going towards the minesweeper, we put out tactical chaff. They saw that and shot against it, from a range of 40 kilometers or something like that. All of a sudden, we were really inside a battle, a missile battle, as we had trained.

Those missiles were huge, like big balls of fire on the horizon coming at you; it looks low on the horizon but it is not that low and it is coming towards you. We all turned against it, performing all the anti-missile tactics we had practiced, including informing headquarters. It was my job to tell headquarters that we had been fired on, and I can imagine how they felt, because it suddenly went quiet, since it takes about two and a half minutes before the missile arrives. In addition, we were not yet in range for our missiles. Nevertheless, it worked, and their missiles missed. Some of them were quite close, some of them shot down, but they all missed. They missed because our tactics worked. We got into range. We even had enough time to say, "You take this one; you take that one," and so forth. By that time, we were already in range.

I remember one case where the second missile we fired did not find its target because it had disappeared. Apparently, the first missile hit a missile that was still on deck, and the vessel exploded. One of the three vessels decided on a different tactic and went straight for the shore. They went aground and the flotilla commander said, "I'm not leaving you here." The vessel that we were onto came as close as it could, as it was very shallow, and from there, it was possible to fire the guns. The commander went up to the bridge to verify for himself and saw it was burning from stem to stern. From navy headquarters, they said, "Okay, enough is enough; it's almost midnight." So we left.

We had excellent intelligence in our preparations and during the skirmishes. It was extremely important that we had a very short chain of command. The flotilla commander spoke directly with the commander of the navy with no go-betweens. Sometimes, when the flotilla commander was occupied with dealing with an attack, I spoke with the commander of the navy.

Udi Erell graduated from the Naval College and served 23 years in the Israeli Navy in various command posts at sea as well as heading the technological naval capabilities branch. His last post was as Commander of the Red Sea Naval Area. He retired with the rank of Navy Captain. Post retirement, he managed two publicly owned companies charged with developing and operating tourism and marine related infrastructure. He founded and manages MatysOnBoard Ltd., which specializes in satellite-based communication solutions for vessels. He is the naval consultant at Israel Aerospace Industries. He holds a degree in economics and business administration from Bar-Ilan University.

2 The Battle of Beirut, 1982

Maj. General (Ret.) Amos Yaron

This is what I can tell you from a division commander's point of view: In 1982, when I moved forward to the northern section, I started the war with one full brigade—the paratroop brigade—and some armored battalions, you could say one brigade plus. At the end, I had eight brigades. Why?

Because the area was so problematic that nobody could support me. I am at the front, you want to give more power, and you want to give me more missions. I need more power, so send me more power. Nobody would come and replace me at the front. Therefore, that was the situation. I finished up with seven or eight brigades and lots of artillery when I actually entered the main built-up area of Beirut.

I was given lots of freedom in 1982 as a division commander. I didn't ask anybody. I did not ask my superiors if I could do this or that. It was my area, and I would do what was necessary to control it. If I needed something from above, it was artillery or air support, those kinds of things. Regarding ideas about how to do it, I did not need anything from anyone.

Maj. General Amos Yaron joined Nahal in 1957. He volunteered for the Airborne Battalion. He later served in several command positions in the Paratroopers Brigade. During the Six-Day War, he served as the operations officer of the reserve 55th Paratroop Brigade that fought in Jerusalem. From 1970 to 1973, Yaron served as the commander of the Nahal Airborne Battalion. In January 1971, he commanded one of the forces that raided Lebanon during Operation Bardas 20, Operation Bardas 54–55, and Operation Spring-of-Youth. In 1975, Yaron was promoted to commander of the Paratroopers Brigade in reserve and later to the regular Paratroopers Brigade. Between 1978 and 1980, Yaron established Division 720 (Judea Division, a reserve division that operated from 1978 and 2004). In 1981, he was appointed to chief officer of infantry and paratroopers.

During the 1982 Lebanon War, Yaron commanded the division landing at the mouth of the Awali River, which fought all the way to Beirut. In 1984, Yaron was promoted to Maj. General. In 1986, Yaron was appointed as military attache to the United States and Canada, a position he held until 1989. Yaron was Israel Defense Forces Maj. General and former head of the Manpower Directorate. He served as the Director General of the Ministry of Defense from 1999 until 2005.

3 | The 35th Paratrooper Brigade in the Battle over Lebanon, 1982

Maj. General (Ret.) Yoram Yair

Since I was part of a separate paratrooper's brigade—the only special brigade in the I.D.F. at the time—when the Northern Command gave the order to all the divisions, I participated in order to hear what my mission was. The idea was that three divisions would enter at the first phase and then a fourth in a different sector, in order to clean up the area, would push out thousands of PLO fighters so that if the Syrians that spread around southern Lebanon interfered, we would push them out as well. This was our border; one division would come from here, a second division from there, and a third one would move along the coast. The paratroopers would land in a very high region relative to our area, with very steep mountains about 1,800 to 2,000 meters high, in order to block the Syrians if they tried to come down or if someone tried to escape from there.

I took on that mission and returned to my headquarters. For a day or two we studied it, and after two or three days we had to present our plans to the Northern Command. I presented my plan, how I would cover the whole area and block it, everything. However, when I finished presenting it before all the division commanders and the staff of the Northern Command, I asked the Northern Command commander—who was a very tough general—if at the end I could speak with him privately in his office since I did not want to embarrass him in front of everyone. He approved of my plan, and, when we finished, we went to his office where he asked me, "Yaya, what do you want?"

I said, "Listen, you gave me a mission. It is okay, I have no problem with it. The mountains are very high, the landscape is beautiful, and the air is wonderful there, I am not worried about anything. But if you try to reach me with your armored division, the roads there are very narrow, and if one tank rolls off its truck, the whole division will be stuck. It is like Switzerland, very steep terrain and you can get dizzy just looking down. I am not sure that your people looking down a stereoscope can get a clear three-dimensional view. I have another idea . . ." It was a lesson learned during a huge operation that we had participated in four years earlier, the Litani Operation, whose objective was to clear the PLO zone. What happened there was that we probably killed about 200 of them, but thousands of others escaped to the north, and, when we evacuated the area, they came back, so nothing was achieved. Therefore, I said to the commander, "I suggest that, instead of positioning me in an area where I have no actual influence, put me somewhere else. All the roads here travel from east to west. The only road—which goes from north to south—is along the coast, and it is the main road. If I make an amphibious landing there, I will block the road so no one can run away and escape from the south, and the Syrian division deployed there will not be able to come down. Okay?" He accepted it.

During the night while I was waiting for the armored division to join me, I studied the map, and for three or four hours, I attempted to find a way to avoid going into this narrow bottleneck in order to exploit my capabilities as special infantry. We were a crack paratrooper's brigade, but this was not a place where I could use all my advantages; quite the contrary. I was sitting by myself with the map, trying to figure it out. Then I found a small, very narrow road climbing up, which could be reached by a circuitous route.

I asked the Northern Command commander for permission to execute my plan. I asked him, "Why would you put all your eggs in one basket? You have forces with overwhelming power. You have one Golani infantry brigade here and another brigade there, and you have so many tanks and artillery. What do you care if I sneak in there, if you let me?" Then he said that I would have to make a detour of 70 kilometers and that it would take me a while, but finally he agreed to allow me to execute my plan.

I always say that if you perceive yourself as open-minded, and if you accept that the officer under you can also come up with new and creative ideas, you will develop open-minded officers under your command. Conversely, you will block this process if you do not encourage officers who are creative, open-minded, and who do not follow conventional lines of thought. Creative thinking is needed today much more than in the past.

I believe that if you feel confident in yourself, you will allow the people under your command to express themselves. If you lack confidence, you will not want anyone to contradict you or shake your authority. It is very important for you to place yourself willingly under criticism. Do not even call it criticism, but being observed by your subordinate officers. When you are commanding a tank with three other crew members, the probability that you are the most intelligent person in the crew is 25 percent. When you are leader of a ten-man squad, it is 10 percent, and if you are a company commander, it is 1 percent, and so on. If you are a division commander, it is as likely as winning the lottery. So remember one thing: in all probability, the higher you go, the more people there will be under your command who are more intelligent than you.

Maj. General Yoram Yair drafted into the I.D.F. in 1963. He volunteered as a paratrooper in the Paratroopers Brigade. In 1964 after completing Officer Candidate School, he returned to the Paratroopers Brigade as a platoon leader. In the Six-Day War he served as a company commander in the Gaza Strip. During the Yom Kippur War, Yair took command of the 50th Paratroop Battalion and led it through the battles in the Golan Heights and in the Sinai Peninsula. Later, he commanded a reserve Paratrooper Brigade and the I.D.F.'s Officer Candidate School. In the 1982 Lebanon War, he led the 35th Paratroopers Brigade during heavy fighting against PLO operatives and the Syrian army. Afterwards, he commanded the Division in counter-guerrilla operations in South Lebanon. From 1992 to 1995, he was the head of the I.D.F.'s Manpower Directorate. Later, he served as Israel's military attache in the United States.

4

The Northern Commander in the Second Lebanon War, 2006[1]

Colonel (Ret.) Boaz Cohen

Once you understand that you are at war, you are supposed to act as if you at war: deploy your forces as if you are at war, use your firepower as if you are at war, and act with a specific set of values that are relevant for war, not for peacekeeping operations and not for limited, Special Forces or raid operations.

I think the basic problem of the I.D.F.'s organization at one point during the Second Lebanon War was that we (and I, of course, was a part of the senior leadership at that point) did not decide that we were at war. The issue of not understanding that you are at war was being responsible for our best officers behaving in a specific way.

A little bit of initiative is, from my point of view, one of the most important values and principles of war. On the one hand, a military organization must be very disciplined; it has to be. On the other hand, if you do not have commanders from the lowest level up to the highest level that are not doing exactly what they are told to do, there is no way you are going to succeed.

General Sharon is a prime example. I do not know if you have heard that in the Yom Kippur War, he did exactly the opposite of what he was told to do. I can give you many further examples. In that war—and, unfortunately, I do not think we have changed much since then—commanders and [soldiers] were trying too hard to obey exactly what they had been ordered to do, and this replaced initiative.

1   Presentation to Australian Art of War visitors, 2015

I will give you another example, again from one of our best brigade commanders. He had the opportunity to cross the Saluki River by Bint Jbeil with no one on the other side.

I think as a brigade commander, I would never ask; I would just do it, and if you want to stop me, stop me. However, he asked for permission. The division commander said that he had to think about it and asked the general of the Northern Command for approval. About two hours beforehand, there had been a discussion with the U.S. government in which Condoleezza Rice had drawn a red line through the Saluki River and claimed that we could not cross at that time.

However, the line was very wide and could be filled by a whole brigade. Therefore, you cross over and explain later. What happened? About five or six days later, we crossed it and paid dearly, with more than twenty soldiers killed. We were facing the problem of understanding that, when at war, you have to act differently. It is not a peacekeeping operation, and it is not some kind of limited confrontation. It is war. When at war, you have to make decisions, and there are some consequences. However, if you are not willing to make them, you cannot be a commander. From my point of view, I think we failed in that respect.

I would like to discuss another topic: C4I systems. It is a very successful program, but you have to understand a few things. Firstly, at least from headquarters, C4I systems give you the illusion that you can control the squad or team from headquarters. This is a mistake. It does not allow you to control anything; rather, it just gives you a better picture, and that is it. It does not replace the fact that a commander has to be at the front and make decisions from the best standpoint over [their] area of responsibility. Nothing has changed due to C4I systems. It has taken time for people to understand that. Although they have the ability to see the front tank or the front squad, and the sniper (the tip of the arrow), they have to be at the front, and they have to command. Nothing has changed on the soldier's level, even after you install a C4I system. It does not change the most important basic principle: the commander has to be at the front.

I think that few people are experienced in full-scale war and understand that the decision-making process, the independence, and maybe the loneliness of the commander in a full-scale operation are completely different from that during a low-intensity conflict. In a low-intensity conflict, you have the whole system supporting and supervising you and monitoring everything, every step of the way. In a full-scale or high-intensity conflict, you are alone; you have to make your own decisions and take the initiative. If you ask too many questions, you will get answers. They will tell you stop, wait, and let them check. If you do not want to hear those answers, do not ask the questions. I think that, in the previous generation, they asked a little bit less.

The issue of values, of growing up in a generation that is willing to take the initiative while nothing has changed in society, is very hard. It is a long process, and I am not sure if it is even possible because people grow up in a specific atmosphere. However, if they do not grow up in an atmosphere that allows them to make mistakes and not get their heads chopped off every time they do so, they will not make mistakes and learn. In fact, they will not do anything at all. They will try to limit their opportunities for making mistakes.

If you grow up in an atmosphere that instead allows you to initiate and make mistakes—and, of course, from my perspective, that is how I try to raise my [subordinates]—that if you have made a mistake and it is a result of negligence, you truly need your head chopped off. However, in any other case, if you made the wrong decision because you did not see the complete picture, it is to be expected. Nowadays, everything is open to the public; anyone who makes a mistake is brought to public trial. I am exaggerating, but that is the situation. In such an atmosphere, you have to explain to the public that it is hard to bring up a generation of officers who are willing to take action. Further, we have excellent officers. The question is how you educate them and what they are willing to take on and what sacrifices they are willing to make. In that war, we had excellent people in every position, but we still made mistakes.

Col. Boaz Cohen recruited to the armored corps and later served as a platoon, company, and battalion commander. In 2001, he was appointed commander of the 188th Brigade. He later served as a military attaché in the United States and was a Northern Command Operation Officer during the Second Lebanon War. Today, he is vice-chairman of land systems at Elbit Corporation.

5

Command and Leadership in Operation "Defensive Shield"[1]

Brig. General (Ret.) Ofeq Bukhris

Introduction

Operation "Defensive Shield" presented a new challenge to I.D.F. commanders due to the need to take over both refugee camps, crowded with inhabitants, and low, dense buildings in order to attack terrorist infrastructures while minimally affecting innocent bystanders and civilian installations. To that, operational difficulty also added the potential price of error, which in certain areas (such as the Mukataa in Ramallah) was liable to involve the I.D.F. and Israel in serious difficulties due to a micro-tactical blunder.

There was an additional—no less serious—worry present in the commanders' minds, especially at the senior level, prior to entering cities and fighting in them; namely, the fear of heavy losses. The I.D.F.'s experience in previous wars in built-up areas was traumatic. The conquest of Jerusalem in the Six-Day War, the conquest of Suez City in the Yom Kippur War, and fighting in the Lebanese cities, especially Western Beirut, [was] engraved on the senior commanders' memories. Furthermore, the events in the tunnel at the Western Wall and the traumatic events surrounding Joseph's Tomb were still fresh and added to apprehensions that loomed in the background of the battle plans, thus magnifying the operational challenge. As in any war, that challenge lay heavily on the shoulders of field commanders at various levels.

---

1   Reprinted with permission from *Ma'arachot*, Vol. 388, 2003, pp. 32–57.

Like any war, this one also faced the commanders with dilemmas: How were they to plan the operation within a critical period? How would they generate commitment to the mission along the entire chain of command? How would they solve the usual dilemma of command at the front as opposed to control from the rear and what would be the correct composition of the forward command in such a crowded battle sector?

These dilemmas [were] further complicated by the lack of a command doctrine suitable to the conditions described above. An additional factor leading to doubts was a lack of experience. The battalion and brigade commanders had accrued much experience fighting against guerrilla forces in Lebanon, but it was clear that fighting in refugee camps was another matter altogether. The commanders lacked experience in leading relatively large forces into refugee camps. In the course of the fighting and from operation to operation we accumulated experience, some of which [is] described below.

In the present article, I have attempted to focus on the leadership aspects of "Defensive Shield" that appear to me significant for future operations. There might possibly be nothing innovative about these lessons, but they are valuable enough to warrant analysis.

The present discussion will focus on command at the brigade level, its connection with the battalion level and the leadership style of the brigade commander, as they appeared to me as commander of the 51st Battalion. I will base my remarks on actual experiences during the war.

The Dilemma

At the brigade level, it is very difficult to control a battle taking place in a refugee camp. The forces are very crowded, fighting shoulder to shoulder, and at times, face to face, with the enemy, while reports from below are not always clear and only provide a partial picture. What is especially lacking is the famous elevated observation point from which the brigade commander can maintain absolute control over the brigade. In refugee camps, no such vantage point exists.

The basic assumption at the brigade level is that the commander maintains control throughout the battle. This might come to expression in the tactical headquarters at the rear or at the forward command, either on foot or by vehicle. From my experience as an officer in the Brigade Operations Branch, I am aware that even in its full capacity the tactical headquarters is not very efficient. During my service as an operations officer, the Golani Brigade commander, Shmuel Zakai, decided during maneuvers and infantry exercises to minimize tactical headquarters from eight to ten people only. The disadvantages of that body magnified tenfold in the refugee camp, in which movement is often limited to single file in narrow alleyways. This means that the command post in its usual composition during maneuvers does not enable the brigade commander to maintain control when he is absent from the operations room. In other words, the brigade commander cannot simultaneously control and command at the front [and] also gain a direct impression of the battlefield, although that is what is demanded of them according to techniques that are practiced at headquarters and during brigade-level exercises.

The brigade commander's major dilemma is where they should be located. The answer to that question depends on the answer to two other ones: Where is it possible to understand better what is happening in order to influence and control the battle? How is it possible to overcome the contradiction between the command doctrine, which demands their presence at the front from time to time, and techniques that demand constant control over the brigade's battle? According to protocol, doctrine should take precedence over technique and brigade commanders [are] forced to find a "creative solution" to this dilemma. Solutions differ from one commander to another. In my opinion, the model developed by the Golani Brigade commander, Moshe "Chico" Tamir, is worthy of study and examination. Of course, this model was personally suited to Chico and his staff of battalion commanders, but I think it includes elements that might [be] applied in other forms of battle and by other commanders. It [is] fully described below.

Creating Commitment to the Mission

Leadership in battle begins a long time before the actual fighting—in daily routine—but it takes its final shape during battle procedure, which is a series of preparatory activities before setting out on the mission. These include studying the mission, defining missions for auxiliary units, concentrating fighters and equipment, preliminary patrols, observation, and more. In the course of the battle procedure, the command groups conduct meetings in which missions [are] defined and the necessary adjustments made. Chico's approach was one of communal leadership, not only regarding responsibility for the mission, but also in forming the plan (not in order to waive responsibility, but to perform the mission in the best possible manner).

After the brigade received its mission, the battle procedure opened between the battalions and the brigade. The assignment of missions was very general in order to allow each battalion commander to determine the maximum achievement possible for his battalion. The brigade's operational concept formed the basis for the communal planning, but the battalion commanders [were] given much freedom in defining their missions. The logic guiding the Golani Brigade commander was simple, although it involved certain risks. The simplicity [was] expressed in each battalion commander defining the mission for [themselves]— in conjunction with the brigade commander—thus creating total commitment to the mission. An additional benefit of this procedure was that the brigade commander did not leave a "hysterical margin" and define only seventy or eighty percent of the battalion's capabilities, but rather left it to the battalion commander to define what his unit could accomplish. Such an approach ensured that the battalion commanders were committed to the mission and would define their unit's abilities closer to one hundred per cent.

The Golani Brigade commander employed that method in countless operations. One of these was the conquest of the Mukataa in Ramallah at the opening phase of the "Defensive Shield" Operation. In that mission, the brigade rushed to the scene in the early morning hours of the Passover holiday and was assigned to conquer Ramallah and occupy the Muqata'a that same night. It was necessary to carry out the battle procedure while collecting the fighters from their homes, concentrating war materials from base and gathering together AFVs from all parts of the country. All these activities took place according to a very limited timetable and had to [be] complete[d] by nightfall. In addition, we understood that the enemy in Ramallah [was] deployed toward our arrival and had sent out "listening posts,"[2] but we did not know where they were located.

In such an impossible situation, a commander's natural tendency is to dictate orders to [their] subordinate officers. However, that is not what the Golani Brigade commander did. We arrived at Central Control Headquarters, each battalion or unit received a general sector of responsibility and a process took place of communally defining missions. After studying the mission as thoroughly as possible under the circumstances prevalent at the time, each battalion commander presented his program to the brigade commander. The final product of that process was a brigade plan suitable to each battalion according to the number of men it had managed to accumulate, their degree of tiredness, and their level of training and experience. In this manner, the brigade commander formulated the mission exactly according to the ability of the units based on an operational concept and stratagem that [was] originally defined. Obviously, it was necessary to make adjustments and fit the pieces of the puzzle together in one integrated picture, which [was] accomplished

2   Nickname for a soldier or group of soldiers that are posted at night outside a defended area in order to listen for suspicious movements and announce imminent danger (From the *IDF Dictionary of Concepts*, 1998).

by a more basic second orders group that also included commanders down to the company level. This is an example of how a brigade can [be] activated in a very limited time while most of its forces are still at home and succeed in its mission by very involving commanders and units. (I doubt if anybody would dare to carry out an actual brigade exercise in similar conditions . . .)

The Second Orders Group

The second orders group represents the high point of combat procedure before a mission. Until that phase, most of the work [is] performed separately in the secondary units, with each unit studying the mission and preparing to execute it. At the second orders group phase, all question marks become exclamation points. The orders group has three main goals and in order to attain them in the best possible way, the forum includes two levels below the command level. They are as follows:

- All the pieces of the puzzle [were] assembled into one complete entity, thus changing the operation from a collection of individual battles to a complete brigade battle. All possible points of friction [were] ironed out among the secondary units. A physical connection [was] formed among all the junior commanders, with the brigade commander having the last word in case of disagreements.

- Each battalion and company commander [was] enabled to stand before the entire brigade forum and state clearly, "My mission is . . . " thus declaring [their] commitment to carrying out the mission. In my opinion, this has great value—especially in times of crisis—for ensuring that the mission will [be] performed under any conditions.

- A forum [was] provided for the brigade commander to explain the principles guiding the operation and indicate the common denominator for all the units in the battle. This was also an opportunity for him to present his ideological and professional credo before dividing and only having contact over the signals equipment.

It is worthwhile broadening the period devoted to the second control group, even in a brief battle procedure. That appears to be one of the major "Defensive Shield" issues [that] arose in the operations. It may be said that the control group played a significant role in creating cooperation among the brigade's various components.

Face-To-Face Meetings with Battalion Commanders

Defining missions communally might involve the risk of a lack of frankness between the battalion commanders and the brigade commander; thus, a situation might arise where the battalion assumes a mission that is too "big for its boots" and the brigade commander is hesitant to point this out. In other words, the battalion commanders might take upon themselves missions that they cannot succeed in due to not wishing to devalue their unit.

The brigade commander overcame this danger in an original way: by means of intimate meetings between himself and the battalion commanders. These took place before the second control group or immediately after it. During the meetings, the brigade commander asked the battalion commanders to express what was bothering them about the plan, what required last-minute changes.

I will mention two examples among many of this technique. I was very worried about the problem of evacuating the wounded during the occupation of the Jenin refugee camp. We had no prior experience in evacuating [the] wounded from refugee camps and it was highly likely that there would be casualties. The battalion commanders raised this matter during those meetings with the brigade commander and reasonable solutions [were] found which we took with us into battle. In fact, two men from my battalion [were] badly wounded. The meetings with Chico had two results: the first was that dividing responsibility between the battalion and the brigade was clear and immediate; the second was that officers who were studying at the time and recruited to the brigade established a brigade rescue team. That team was set up ad hoc because of the meetings between the brigade commander and the battalion commanders and thanks to it, one of my wounded men survived.

An additional example of the results of those meetings took place before the occupation of Hebron (after the attack at Adora). A brief but logistically complicated battle procedure was conducted before that operation and toward its completion, I had the feeling that something was missing and that we were lacking in preparation. After the second command group, we met again and I was not the only one who expressed those feelings. The importance of those meetings was that sharing what we saw as problems with the brigade commander and analyzing them brought about a better understanding of the missions and their various phases. During the operation, a successful common language [was] formed that was significant for directing the battle and planning its development. There might be readers that think that such a state of unpreparedness should already disappear during the battle procedure. However, in the almost impossible timetables we were working with, the cooperative method of constructing the plan saved considerable time. The forum created by the brigade commander, which provided solutions to a complicated reality, was an efficient way of involving the brigade in the missions, despite the short time span at our disposal.

During the operations, there were many examples of open, personal meetings, whose value became evident. As far as I can remember, these encounters did not result in changing the basic plans, but made it possible for us, the battalion commanders, to present the brigade commander with what we saw as weak spots. Their practical outcome was that when directing the battle, the brigade commander was aware of what was troubling the units, allowing him to pay attention to those weak spots, especially when they led to complications.

The method of consulting with the battalion commanders was also relevant in the midst of battle. The event that best personifies the advantages of that method was the occupation of the Tulkarm refugee camp. The mission assigned to the 51st Battalion was to conquer the city of Tulkarm and form the reserve for occupying the refugee camp (much to my consternation as its commander). After occupying the city and before taking over the refugee camp, it was difficult to advance as planned. At that point, the brigade commander, Chico, consulted with the 12th Battalion commander, Oren, who [was] situated north of the camp and [was] attempting to break in with his battalion. The brigade commander also consulted with the commander of the reconnaissance battalion that [was] especially formed ad hoc for the mission and was attacking from the south, deliberating with both of them how to proceed. Afterwards, the brigade commander liaised with my forward command (the 51st Battalion) and decided there how to attack further and what the units' missions would be. The renewed attack on the refugee camp, this time from another direction, surprised the terrorists and resulted in their defenses colonel lapsing and their surrendering in less than twenty-four hours. This rapid surrender [was] achieved—among other things—thanks to consultations between the brigade and battalion commanders. The brigade commander understood that he could not decide everything on his own and rely on reports, but only on first-hand impressions.

Consultations between the brigade commander and the battalion commanders took place while directing the battle by means of encounters in the field in which the brigade commander attached himself to the battalion commander's forward command or by means of the signals network. Meeting the commanders in the field was the preferred method.

Of course, the battalion commander's opinion was not always accepted and general considerations—as understood by the brigade commander—prevailed, but one thing was clear: orders [were] based on reality and not issued as detached commands from above.

Command from Within

Clearly, the brigade's occupation of the Tulkarm refugee camp would not have succeeded so convincingly without Chico's presence at the battlefront. The involvement of another battalion in the battle within the camp, which was a significant step, only took place after the brigade commander's visit to the field and his weighing the advantages and limitations of such a step. Furthermore, the decision to allow the terrorists to surrender instead of continuing with house-to-house fighting mainly stemmed from his direct impressions of the situation in the field.

In my opinion, it is very difficult for a battalion commander to describe the situation to the brigade commander over the wireless and it is even harder to convince him how to proceed when he does not experience the reality of the battlefield. The Golani Brigade commander actually commanded from within, thus succeeding in bringing about the refugee camps surrender relatively quickly and with minimal casualties.

Leadership in battle by internal command is of prime importance—especially before significant decisions or after crisis points. The brigade commander's presence on the front lines and [their] direct impressions—even if partial—afforded [them] some advantages.

The first of these was a better understanding of the operational problem; this, as stressed above, had repercussions for directing the fighting.

The second advantage was [their] ability to experience personally the battlefield: the enemy fire, the apprehension, and the fear of casualties. That direct experience enabled him to be part of the battle and allowed us, the battalion commanders, to know that the man giving the orders understood about what he was talking.

The third advantage was that the brigade commander's presence among the fighters [was] on the front lines; that afforded them a sense of security, for two reasons. The first was that they knew that the brigade commander was with them and would not send them to "crash into a wall" and the second was that the brigade commander's presence imparted the sense that things were under control.

Internal Control at Jenin

During Operation "Defensive Shield," I also participated in the Battle of Jenin under the command of the commander of the reserve infantry brigade. In that battle as well, the brigade commander had to decide how to command the forces. The situation at Jenin was particularly complex—among other things, since the brigade's units were not organic and in the course of nine days' fighting, new units were added. The method selected by the brigade commander in order to overcome this dilemma was to command from the brigade's tactical headquarters in order to maintain the flow of control, but at the cost of commanding from within.

In my view, that decision was a doctrinal error, as command doctrine demands that the brigade commander command from within for at least some of the battles. The brigade level is a tactical one and as such, its headquarters cannot rely on reports only. This principle is even more valid when the battle continues for more than a 24-hour period. It appears that the technical difficulties stemming from control methods employed by infantry brigades in the form of tactical headquarters are a barrier making it difficult for the brigade commander to command from within. In face of this reality, the question might [be] asked if tactical headquarters fulfils their purpose in their present form.

In my opinion, especially since the brigade was so heterogeneous, the brigade commander should have definitely commanded from within. When a brigade is organic, the commander is better acquainted with the battalion commanders' reports and can read between the lines in order to get their message. In a brigade where the battalion commanders are strangers to the brigade commander and to one another, the only way to obtain correct information is face to face, in addition to long-distance reports. When the commander's objective is to gain control, being in the rear makes it possible to only partially achieve this due to failures in communication that stem—among other things—from a lack of a common language among senior and junior commanders.

The most vital component to damage by the brigade commander's absence from the front is trust. In an organic formation, trust gradually established in routine times and on the battlefield. The result is mutual understanding, which the brigade commander must strive to maintain. However, in a brigade that [is] made up of units that were not present in the second command group, but only joined on during the fighting, the various commanders might not even know their neighbors' names. In such a situation, the brigade commander must make an effort to build trust and if [they are] physically absent from the front lines, a sense of distance and a lack of cohesion are created that will seriously curtail [their] ability to lead the forces. A comparison between the Golani Brigade and the brigade that fought at Jenin indicates that in the latter, the cutting conditions of the latter were better, but the content of the commands and the connection with them in reality were much superior in the Golani Brigade.

The most striking example of a brigade commander's internal command was actually when the battalion was not under the command of the Golani Brigade. After six days of fighting in the Jenin camp, the Golani Brigade commander paid a visit to my battalion inside the refugee camp. He was not responsible for either the sector or the operation; however, his presence in the difficult conditions prevalent at the camp imparted a sense of security to officers and men alike. I am convinced that such an act of leadership contributed to the battalion soldiers' morale at a difficult phase in the fighting.

At the same time, it must not [be] forgotten that internal command on the brigade level has some weak points. The first problem is how the brigade commander is supposed to arrive at the company commanders' forward command or visit a particular company without breaking the chain of command. However, this problem is not insoluble. For example, the Golani commander ensured during his visits that his contact with the fighters would be in the sphere of leadership and would not involve the issuing of direct instructions. Instructions given to the battalion commander as a warning command and afterwards the brigade operational branch transferred the command in an orderly manner in order to avoid coordination problems. In fact, the brigade commander's visit chiefly [was] aimed at receiving impressions of the situation at the front and deriving operational insights regarding the entire brigade, not as a means of pulling rank.

An additional possible weak point is reduced control over the forces while the brigade commander [is] occupied with commanding from within. This can [be] solved in two ways, which are mutually complementary:

1. Exactly for this purpose, the brigade has a second-in-command, a chief staff officer, and an operations officer. While the brigade commander fulfils [their] duties as internal commander, control remains at headquarters and [is] carried out by those officers.

2. An additional solution is simply to trust junior officers to carry out their responsibilities. However, this demands that the brigade commander have faith in them and make this known to them. A commander who shuts [themselves] up in headquarters for fear of losing control transmits a lack of trust in his junior officers. It is doubtful if there is any value in control of that type.

Two aspects of leadership were evident in the Golani Brigade commander during Operation "Protective Shield":

1. He created true cooperation between battalion and brigade and a mutual obligation to carry out the operation; as a result, his orders were applicable and based on reality.

2. Internal command served to create a sense of comradeship during battle, especially at moments of crisis and before significant turning points.

The following conclusions may [be] draw[n] from the above:

1. During exercises at the Brigade Headquarters Training Center, the brigades must learn ways of formulating battle plans together with the battalions in a speedy and efficient manner (in addition to, not instead of, existent techniques).

2. The infantry brigade commander's tactical headquarters must [be] adjusted to commanding infantry in crowded, closed-in areas that are not suitable to control from an elevated observation point.

3. During exercises at the Brigade Headquarters Training Center, emphasis will [be] placed on the need for internal command and it will form a significant component in training exercises.

Summary

In the course of Operation "Defensive Shield," commanders were required to command and control battles in a complex operational reality.

The dilemma they had to deal with was whether to be present with junior commanders and fighters in the field in order to form direct impressions of the battlefield (at the cost of losing control due to leaving the tactical headquarters).

To maintain absolute control over the forces from headquarters (at the cost of losing direct contact with the field).

It appears that those choosing to command from headquarters for fear of losing control over events cutting themselves from what is happening in their sector, do so due to their basic mistrust of the framework they are commanding.

Commanders who believe in their subordinates have no fear of temporarily losing control (and for that purpose, appoint a second-in-command) and allow themselves to perform the most significant act of leadership, namely, maintaining direct contact with subordinates and the reality that they are facing. In the battles in which I participated, I felt trust in our ability to carry out our missions, which [was] expressed, among other things, by the commander's willingness not to decide everything by himself and his basic belief that his soldiers would follow him into danger and would succeed in overcoming difficult situations.

In my view, when those commanding forces trust them, share their hard times, and do not shut themselves away in headquarters, they are capable of facing challenges that a short time previously seemed impossible of attainment.

Brig. General Ofeq Buchris recruited to the I.D.F. in 1988 to the Golani Brigade (Infantry). He served as Company Commander, Battalion Commander, and was involved in many operations, including Defense Shield in Jenin. In 2010, he was appointed as the Golani Brigade Commander. Later he was appointed as Division Commander and the Command and Staff College. He retired from the I.D.F. in 2016. He is a graduate of the U.S. Army War College. He holds a B.A. and an M.A. from the Hebrew University Jerusalem.

6 | Junior Command in the Gaza Strip[1]

Major Vered Vonokor-Hai and Lt. Colonel Yotam Amitai

Regarding platoon or even squadron commanders, it is clear that they are responsible for most battle action in the field and they have the greatest influence on outcomes. Official tactical doctrine claims that the direct commander of the unit in the field is the best judge of the situation and that [they] must give a large degree of autonomy. The constant presence of senior commanders is liable to obstruct decision-making and remove responsibility for the battle from the lower tactical levels.

The location and circumstances of current conflicts have increased the debate about the dictates of reality and junior commanders' independent functioning in the field.

This may [be] demonstrated by the comments of the former commander of the Gaza Division, Maj. General Israel Ziv:

> A significant aspect of senior commanders' behavior is reliance on their subordinates . . . this is due to the fact that the operational level will always be more expert in what is happening than their immediate superiors. But there is tension between the desire of senior officers to maintain centralized control and minimize possible mistakes on the part of forces in the field.

Nowadays, the senior commanders in the Gaza Strip whom we interviewed were convinced that this turbulent sector is the most

1   Reprinted from *Ma'arachot*, Vol. 389, 2003.

difficult and exhausting in current warfare. A rising number of incidents alongside the characteristics of fighting in the sector (mainly involving static fighting) makes the role of junior officers extremely demanding on the one hand and heavily responsible on the other.

One senior commander fighting in Gaza stated: "Today the battlefield is the province of platoon commanders since they are the first to respond to incidents and are immediately required to come up with solutions, while most of the time we, their commanders, arrive when it is all over."

A battalion commander added:

> In this crazy sequence of events, my role is first and foremost to prepare the forces, guide and educate them, but in real time when something happens, it is unlikely that I will be there. The person who has to come up with the answers is the most senior commander in the area, who is in fact a junior officer.

Examining attitudes of junior commanders toward their seniors' intervention in the sector reveals differences between company commanders and lower ranks. For example, platoon and squadron commanders are generally grateful for supportive intervention on the part of their superiors. They believe that their greater experience makes a significant contribution and that their intervention enables a quick response due to circumventing the chain of command. One platoon commander said

> When there is a real-time incident, the battalion commander jumps in over everybody else. I personally believe that the battalion commander is better than me . . . the brigade commander also arrives to command from close up and this is important . . . because if one goes according to the chain of command, it lengthens the process, creates confusion and misses opportunities. Thus at the time of the incident the senior commander must be the one to make the decisions.

Unlike squadron and platoon commanders, most company commanders claimed that their knowledge of the area and their own men gave them a considerable advantage and that they should give more freedom in managing operations. One of them said, "Intervention on the part of senior officers significantly damages leadership at the junior level, which must ultimately take responsibility for the consequences. When they are removed from the command loop during action, their sense of responsibility weakens."

It also emerged from company commanders' comments that they viewed senior commanders' interventions as stemming from the great pressure exerted on them and the huge responsibility on their shoulders. One of them described it thus: "The less room there is for error, the less freedom of action there will be."

Major Vered Vinokor-Chai is Organizational Consultant at the I.D.F. Behavior Sciences Department.

Lt. Colonel (Ret.) Yotam Amitay is Organizational Consultant who serves in the I.D.F. Behavior Sciences Department. His last assignment was the Head of Senior Officers assessment branch.

7 An Infantry Battalion Commander in Lebanon, 2006[1]

Colonel Avi Dahan

Uncertainty on the battlefield in Lebanon demonstrated to me more than anything else, in war, it is impossible to achieve full security and scientific accuracy. When I attempted to achieve these things (for they are a part of human nature), I hesitated too much and the fighting passed me by. In other words: I gave up vital speed and agility—especially against an army working under guerilla conditions like Hezbollah—in order to obtain a clearer, more exact picture.

In time, my junior commanders accrued more and more experience and self-confidence, which brought them to a better level of professionalism and operational efficiency. Because of this improvement, I could decentralize more authority to company commanders. Furthermore, commanders would change their plans in the midst of battle in order to perform their missions in full cooperation between their forces and fire support on land and air.

The investigation that [was] performed later revealed that the freedom of movement that [was] given to company commanders during tough hard battles created the operational conditions to continue the attack. If this had not been the case, and if the method during the battle had been command and control, the commanders would have had to halt at every phase and wait for orders for their next move; this would have given the terrorists precious time to redeploy, reattach, and win the battle.

1 Reprinted from *Ma'arachot*, Vol. 409–410, 2006.

The commanders learned from me what was necessary to accomplish (the mission) and they planned and decided on how to do it (the method). In other words, mission command and the fact that company commanders understood the battlefield and sought contact enabled continuous fighting that resulted in the eradication of eight terrorists occupying the territory dominating the village and completing the mission.

The core of this command philosophy is apportionment of authority to the lowest command levels—company, platoon, and even squadron commanders. The company commanders were assigned missions and a wide margin of freedom to make decisions while carrying out the mission, but they made sure to keep me informed as commander of the sector.

One of my major insights from the Lebanon War was that I could never totally understand what was happening in the battle and have complete control of it, especially if it took place in an urban setting or on difficult terrain against a guerrilla army. Thus, I had to rely on my subordinates to carry out my orders as I would. In addition, I learned that the degree of faith that I placed in my juniors—company and platoon commanders—was a source of power.

The more leeway I granted them to think and act, the more their initiative, independence, and their will to make contact with the enemy increased. In this way, I enabled them to develop and flourish on the battlefield and gained valuable time to devote to my major responsibilities as commander: managing the battalion's fighting, activating enveloping forces, activating the battalion's reserves, and coordinating among the fighting forces.

It is important to note that only toward the end of the war could I allow myself to apply this method, as mission command demands training and combined operations at the company and battalion levels. In fact, in Judea and Samaria, we did carry out battalion-level operations and made sure to conduct platoon-level training exercises, but the battalion participated in a very different type of fighting than that, which took place in Lebanon.

From this stemmed one of my chief insights from the Lebanon War: I will never be capable of totally understanding what is happening on the battlefield and have full control over it, especially regarding fighting in urban areas or complex sectors against an army employing guerrilla tactics, while succeeding to command the battalion in a professional manner.

The battle, the frequent clashes, and the considerable casualties (two dead and thirty-two wounded) on our side were important variables that affected decision-making. Above all, I attempted to apply the mission command philosophy in that battle. We attacked the fortified village on two sides of its main street.

At the commanders' meeting that [was] held the night before the attack on the town of Qantara [Lebanon], I emphasized to the company commanders the lines of advance of each force. I set a minimal number of conditions for their movement in the village and enabled free discussion among them in order to coordinate covering fire and the forces' position during the battle.

It opened with strong resistance on the part of the terrorists and within twenty minutes the battalion had suffered, one dead and twenty wounded. The company commanders continued advancing and seeking contact since they understood the battalion's operational concept.

The operational conclusion: In order to prepare a whole battalion, it must train together.

Only in that way will it be possible to strengthen commanders' professionalism on all levels, increase their self-confidence—in their abilities and in the war materials in their possession—and, of course, strengthen the senior commander's faith in company and platoon commanders. Von Manshtein, one of the most talented Wehrmacht commanders in the Second World War, wrote about it in his book, *Lost Victories:*

One of the sources of power in the German command, which made it unique, was the amount of trust it invested in officers at all levels and the encouragement they were given to assume responsibility, take the initiative and act independently. The senior officer mentions in his orders what the mission is and its general direction and leaves it to his juniors to decide how to carry it out, while providing them with the best possible resources: training, weaponry and doctrinal knowledge so that they will successfully complete the mission (1998).

These, then, are the advantages of mission command: improved motivation, initiative, and creative thinking at all levels; the battalion commander's responsibility to develop and nurture [their] juniors' initiative, trust their strength, and develop their professional independence, enabling them to make quick decisions on the front lines; exploit every opportunity; and meet the challenges presented to them, while saving precious time.

However, mission command also has its disadvantages: it places a burden on the battalion commander to encourage shared training maneuvers among company and platoon commanders and to clarify [their] operational, professional, and even moral expectations. At times, junior commanders' decisions are not in keeping with their superiors. For example, parallel to awarding four Medals of Honor for bravery in the war, a platoon commander [was] dismissed for retreating instead of attacking and not fulfilling my expectations of him and the standards of a commander on the battlefield. Commanders' lack of proper training and professional knowledge can result in their making bad decisions that can have a decisive effect on results on the battlefield.

The solution to these two deficiencies is to train forces and prepare them, develop their professional and historical knowledge, and conduct numerous tactical exercises that will contribute enormously to their professional level.

Colonel Avi Dahan recruited to the I.D.F. in 1993 and served as Infantry company commander as well as Combat Engineers Commander, was Battalion Commander during the Second Lebanon War, was the I.D.F. Liaison officer with the USMC, then Brigade Commander, and DY Division Commander. He graduated from the I.D.F. National Defense College.

8 From Surprise to Knockout
The Battle of Wadi Mabuk

Brig. General Gideon Avidor

A major battle during the Yom Kippur War (1973) happened on October 14. On that day, the Egyptian army launched an armor attack— an attack that the I.D.F. was waiting for from the start of the war.

Division 252nd was responsible for the Southern Sector facing the two Egyptian Infantry reinforced divisions, the 7th and the 19th. An armored division from the Egyptian second echelon, the 4th Armored Division, carried out the attack in that sector.

252nd Division assess that the attack will take place in its center or northern axis and deploy its forces accordingly; in the morning, the Egyptians attack on the Southern axis. The Division Commander with two Armored brigades waited at the Division's northern sector; it was a surprise.

The division staff at the Forward Command Post (F.C.P.) took charge, ran an improvised unified battle, and, in five hours, destroyed the Egyptian Third Armored Brigade (Reinforced).

The forces that took part in this battle were these battalions (each from a different brigade): 202nd paratroopers, 46th Armor, 89th Mechanized Infantry, Artillery Regiment (three battalions), and Air Force support.

All that time the Division commander and two-armored brigade waited elsewhere for what was thought be the Egyptian main attack. Later, another Egyptian brigade (the 11th Bde) launched an attack there.

One can see the differences in command reaction to an emergency if it is compared the I.D.F. mission command actions to the Egyptian's looking up for rescue, as discussed by Lt. General Wasl's report.

Introduction

During the Battle of Wadi Mabuk, the fighting in the canal sector did not resemble anything known or planned by the I.D.F. beforehand. The enemy fought in a manner unknown to us previously. The enemy's exploitation of the area was done in such a way as to be unexpected. Furthermore, the forces and materials at the I.D.F.'s disposal were unsuitable to such fighting conditions.

In previous maneuvers and war games, the General Staff assumed that an Egyptian crossing would be their opening gambit. Still, nobody took it seriously enough to think it demanded a response, whether technical, organizational, or doctrinal, or that the divisional or General Staff level would be called upon in the future to cope with such a situation. The I.D.F. did not practice defense doctrine, and what was practiced had never been practiced or adapted to actual, real-life situations.

Solutions had to be created during fighting in the field, by trial and error.

On the tactical level, the tank commanders were those who had to face anti-tank threats with evasive maneuvers and fire management, particularly of the isolated tank, and partially in the tactical framework of including battle intelligence (observation) in firing battles and integrating infantry and artillery in mobile battles.

The Egyptians decided to attack this sector with an armored brigade reinforced by a Mechanized Infantry battalion. The attack was commanded by the Fourth Armored Division commander, who crossed over from the western bank to the assembly areas on the eastern bank the previous day together with the brigade and the battalion. The Egyptians also attacked in the division's northern sector, that of the 164th Brigade, on the "Poreret" Road with the 11th Brigade reinforced by a tank battalion from the 25th Armored Brigade.

The 252nd Division, in whose sector the attack took place, was aware of the imminent attack and deployed to face it. The division expected that the armored division would attack on one of the main roads/axis in the sector, the Mitle Road/Axis, the GIDI Axis, or both of them.

The division saw the "Yore" Road as securing the flank; from the outbreak of the war, there was no significant war activity there. When the 202nd Paratrooper Battalion entered the divisional area, it took positions at "Naveh" Junction to secure the flank and protect the sector from Egyptian infiltration.

The attack came as a surprise to the division. The Division commander and three brigades' commanders deployed to block the anticipated Egyptian attack in the central and northern divisional sectors; a battle developed in the southern area under the command of the divisional operations office.

At that area, a battle developed in which three maneuvering battalions took part (the 202nd, the 46th, and the 89th), each from a different brigade, an Artillery Regiment (the 209th) with three artillery battalions, air support (69 sorties) and logistic support (the 401st Brigade).

The battle was improvised, hastily managed, and fought on the move; it was division-level regarding command but battalion-level in the fighting sector, starting with a surprise and ending with a "knock-out."

The Course of the Battle

Operational background of the fight in the 252nd Division's southern sector is as follows.

The 252nd Division's area extended from the center of the Bitter Lake to three km south of the "Yore" Road.

South of the divisional sector was the "Marshal" area. The closest force in that sector was the 35th Brigade that was stationed at Ras Sudar and other small troops deployed in the area, especially opposite Ayun Musa, from which the Egyptians attempted from time to time to move southwards in the direction of Ras Sudar and Abu Rhodes.

The Division's Mission

The division's mission was to prevent the Egyptian Army from breaking to the east. Our positions were as far west as possible, without entering into massive, exhaustive battles. We maintained a "long-range" firing position to retain contact and place continued pressure on the Egyptians.

The divisional battle order included

- The 401st Armored Brigade commanded by Col. Dan Shomron,
- The 164th Armored Brigade commanded by Col. Avraham Bar'am,
- The 875th Mechanized Infantry Brigade (two Mechanized Infantry battalions and a Sherman tank battalion) commanded by Col. Aryeh Dayan,
- The 209th Artillery Regiment commanded by Col. Ya'akov Erez, and
- The "Bishof " Force (an infantry brigade) commanded by Col. Haim Binyamini.

The size of the forces did not enable a continued presence along the whole width of the sector. Some of the areas were only covered by observations. We planned to divert forces from areas that were not defined as essential or dominating.

The division deployed its forces with two armored brigades forward and a Mechanized Infantry brigade in reserve, a Mechanized Infantry battalion dug in in the Gidi Junction exit area, and a Mechanized Infantry battalion dug in in the Akavish area.

Two of the 875th Brigade's Mechanized Infantry battalions were deployed, one at the exit of the Yore Road and the other at the exit of the "Gidi" Road; since the 35th Paratroopers Brigade were no longer flown to Ras Sudar, the 202nd Paratrooper Battalion was assigned to the division's command on October 11.

When there was Egyptian activity at the Naveh-Yore Junction, the division commander decided to reinforce that flank, making it a divisional sector. It was meant to reinforce the tank company that was active in the sector and release the 401st Brigade to concentrate on fighting on the "Atifa" Road, which was considered the main Egyptian attack road.

The 252nd Division and Its Deployment Towards the Attack
The Headquarters and the Forward Command Team (F.C.T.)

With the outbreak of the war, the 252nd Division was in charge of the entire Suez Canal front under the Southern Command. The Division headquarters was in Refidim with the transfer of command of the Northern sector to the 162nd Division and the Central sector to the 143rd Division. On October 7, in the afternoon, the division sent an F.C.T. to the southern sector, to the western entrance to the Mitla Pass.

The F.C.T. was a mobile body that broke off from the main headquarters. The structure of the F.C.T. was based on five M113 A.P.C.s (command, artillery, intelligence, communication, security) and two 4-by-4 vehicles for the Division commander and the Artillery Regiment commander.

The team consisted of Maj. General Avraham (Albert) Mendler, the Division Operations officer (G3); Lt. Colonel Gideon Avidor, the Division Artillery commander; Col. Ya'akov Erez, the Division intelligence officer (G2); Lt. Colonel Yossi Tayar-Tamir; and the Division signals officer, Lt. Colonel Ya'en Vered as well as a few other officers and NCOs.

Managing the Division Battle

During daytime hours, the division commander met up with the brigade commanders in their forward positions using A.P.C.s and one 4-by-4 vehicle, occasionally with the addition of a helicopter, while the ongoing battle management was performed by the F.C.T. from the Mitla position.

Every day, with the division commander's return from the field to the F.C.T., generally at nightfall, a situation assessment was made, and decisions were made regarding the next day's fighting procedure and those to follow.

The Deployment on the "Yore" Road

The 202nd Battalion (a paratrooper battalion mounting on M3 half-tracks) was assigned a sector as "flank security." Its mission was to occupy the Naveh-Yore Junction and hold onto it, "to keep an eye" on the sector, and to prevent the Egyptians from taking over the Artillery Road ("Nave" Road).

The 89th Mechanized Infantry Battalion's Deployment

The 89th Battalion commanded by Lt. Colonel Yitzhak Shoshani included two Mechanized Infantry companies, "Z" and "H". The battalion deployed in the "Akbarosh" enclosure on the "Izavon" Road; its mission was blocking the road to the southwest. Its heavy weaponry was an 81mm mortar platoon (four barrels) and three 20mm guns mounted on M3 half-tracks.

October 13–A Dark Day

October 13 was one of the division's most difficult days. After a failed attempt to rescue the men from the "Pier" stronghold on the Suez Canal line on October 11 by sea, there was no way to reach them. On the morning of October 13, it was decided that they would surrender in an orderly fashion with the intervention of the Red Cross. The final decision was made at 1000 hours; afterward, the division commander left the headquarters for a tour of the 164th Brigade. His A.P.C. was sent to the "Poreret" Road, and he arrived there by helicopter.

At that time, the Egyptian 11th Mechanized Infantry Brigade attacked the sector supported by artillery; in the artillery attack, the division commander's vehicle was hit and Maj. General Albert Mendler killed.

He was replaced by Maj. General Kalman Magen, who arrived at the division command post around 1800 hours in time for preparation for the next day's anticipated battle. The Southern Command said that tomorrow (October 14) an attack by the Egyptian Fourth Armored Division is expected and alerted the Division.

Upon his arrival, Maj. General Kalman Magen conducted a situation assessment. It estimated that tomorrow's battle would take place on the "Poreret" Road and that the 164th and 401st Brigades would be deployed towards it. On October 14 at dawn, the division commander joined the 401st Brigade in anticipation of the coming battle.

October 14—The Day of the Battle

On October 14, the order of the division force was as follows:

- The 401st Brigade (53 M60 tanks) three battalions: the 195th, the 52nd, and the 49th.
- The 164th Brigade (52 Centurion tanks) two battalions: the 104th and the 106th.
- The 875th Brigade (30 Sherman tanks): one tank battalion (the 128th) and two Mechanized Infantry battalions (the 89th and the 186th).
- The 202nd Battalion (seven M60 and Centurion tanks): two paratrooper companies and a Support company.
- The 209th Artillery Regiment: five artillery battalions, one 120 mm mortar battalion.
- The "Bishof" Force, an infantry battalion from the Officers' School.

The Egyptian Plan

In *The Arab Israeli Conflict*, Lt. General Abd al-Muneim Wasl, the Third Army commander, discussed that battle:

According to President Anwar al-Sadat's instructions, the High Command of the Armed Forces decided on October 12, 1973 to develop an attack eastwards and pressure the I.D.F. to force them to ease the I.D.F. pressure on the Syrian forces on the Golan . . . On the eve of October 12, the orders arrived to develop the attack towards the mountain passes; the hour of the attack was set to the October 14 morning.[2]

The Third Army's Battle Plan

The Third Army's situation at 1800 hours on October 12, 1973 was as follows:

The first echelon of the army east of the Canal—on the Al-Shat-Mitle Road—had 149 tanks fit for battle:

- The Seven Infantry Division (commanded by Brig. General Ahmad Badawi)
- The 19th Infantry Division (commanded by Brig. General Yusuf Afifi Mohammed)

The second echelon of the army west of the Canal had:

- The Fourth Armored Division (commanded by Brig. General Abd Al Aziz Kabil)
- The Six Mechanized Infantry Division (commanded by Brig. General Mohammad Abul Fath Muharram).

I created a plan to develop the attack towards the mountain passes in the Third Army's sector as follows:

Leave the 19th Division's bridgehead unchanged and to attack with the Four Armored Division, from the army's second echelon, which until that time had been located west of the Suez Canal and had not participated in the fighting.

The Third Armored Brigade would mount an attack from the 19th Infantry Division's bridgehead.

The 339th Mechanized Infantry Battalion would mount an attack from the 19th Infantry Division's bridgehead, on the right flank of the Third Armored Brigade through Wadi Mabuk to reach the Mitla Pass.

The 11th Mechanized Infantry Brigade from the Seven Mecha-

2   From the memories of Lt. General Abd al-Muneim Maktaba al-Suruq al-Dawliyah, Cairo (2000) pp. 216–222 (Arabic).

nized Infantry Division would mount its attack from the Seven Infantry Division's bridgehead towards the Gidi Pass.

The Battle from an Egyptian Viewpoint

Maj. General Wasel, the Egyptian Third Army commander, described the battle as follows:

> For fifteen minutes, the artillery bombarded visible enemy targets in the sector and on the advance line flanks.
>
> The brigade mounted its attack, as well as the 339th Mechanized Infantry Battalion.
>
> Despite all the preparations for the attack, the Third Armored Brigade did not succeed in carrying it out, although it reached a distance of seven km west of the Mitle Pass. Having left the range of air cover, it entered an enemy anti-tank ambush; thus, it suffered many enemy air attacks.

In his book *The War Battles on the Egyptian Front* (Dar al-Sharuq, Cairo (2002), pp. 252-258 (Arabic)), Maj. General Jamal Hammad described the battle as follows:

> An airstrike was supposed to take place at the opening of the attack but did not materialize since contact [was] not made with the ground-to-air missile units that transferred to the eastern bank of the Canal before first light on 14 October. Thus, the brigade remained without air cover or air defense.
>
> At the appointed time, the artillery units opened a 15-minute bombardment to secure the brigade's movement, but it did not have any effect since there were no defined targets, and intelligence was very scarce.
>
> At 0630 on the morning of 14 October, the brigade's units began

crossing the opening line from the 19th Infantry Division's bridgehead, under the command of the Fourth Armored Division's Commander and his command team. The brigade advanced in two echelons; the first made up of two tank battalions reinforced by two Mechanized Infantry companies. The second included a tank battalion minus one company. The brigade's reserve included its Mechanized Infantry battalion minus two platoons.

The Israeli Viewpoint, 0600-0800

The attack took the division by surprise. The sector was not considered vulnerable, and the force stationed there was a paratrooper unit meant to secure the flank and not a force that was capable of blocking Egyptian armored attacks.

On October 14 at 0635 hours, the F.C.T. received a warning from the 401st Brigade's intelligence that Observation Post 215 "F" had sighted vehicle movement from the area south of "Polygon," east of Wadi Mabuk. The F.C.T. checked with the 401st Brigade's observation posts (on the Karat Murah ridge) and with the 202nd Battalion and discovered that they had seen nothing. The 215 "F" observation post continued its warning, and, at 0650 hours, the 202nd Battalion's "C" Company reported that tanks were passing by it and entering the Wadi; a "general alarm" was sounded.

The 202nd Battalion CO, Lt. Colonel Doron Rubin, thus described the events:

> At 0650, as I am walking around, I see that they are signaling me to return immediately. I run back and hear Israel Meir reporting to me that tanks are moving from the north in his direction in big numbers and he advises me to fold.

Maj. Israel Meir, who commanded the paratrooper company west of

"Naveh 71," reported:

> In the morning, I picked up my binoculars and didn't see anything. Total silence. Suddenly the entire desert began moving. I saw tanks on 'Naveh' breaking southwards directly at us.

The 202nd Battalion activated the effective power it had (seven tanks, five recoilless guns, and eight mortars) in an attempt to block the Egyptian advance but without result; they passed by the battalion and penetrated deeply into the area.

These reports alerted the division. At that hour, the 46th Battalion of the 401st Brigade was commanded by Maj. David Shoval and positioned at the Mitla Junction. They were in the process of reorganizing as follows: tanks serviced, soldiers showering, phoning home, and performing such activities as were done between battles.

The division ordered the 401st Brigade's commander to send out immediately a tank company towards "Izavon 62," the meeting point between Wadi Mabuk and the road.

46th Battalion's commander, Lt. Colonel David Shoval, described it thus:

> Around 0630, we began hearing over the radio network about clouds of—we understood that something was happening. We had planned on a day of reorganization. I told Yaron ("C" Tank Company CO) to close down everything and prepare the tanks to move out, and a quarter of an hour later, I ordered him to proceed to the ridge dominating Wadi Mabuk south of "Yichil."
>
> It took him 20 minutes to organize; I went out with him and I told Ira ("B" Tank Company CO) to arrange to move out. I advised Col. Dan Shomron (the 401st brigade CO) that I was sending Yaron and Ira as well to deploy north of Wadi Mabuk and I am going with them.

Yaron started moving out at around 0700 hours, Ira about 30 minutes later. I moved out with Yaron, and within half an hour of traveling we were in position, I was about one or two km behind Yaron, who opened fire before I arrived.

Lt. Yaron Ram, "C" Company commander reported that

The entire battalion stood under camouflage nets in the day lager area. At about 0700 hours, we were told that Egyptian armor had been sighted.

The instructions were to move to Wadi Mabuk because the Egyptians were advancing quickly through the Wadi. The G3, the division's operations officer, entered my radio net and began asking what was happening. We arrived at "Izavon 62," where we could already see lots of dust from the area near "point 143."

Lt. Colonel Doron Rubin, the 202nd Battalion commander, described the following:

I took three tanks and sent them above Wadi Mabuk to go and knock them out from the flank; I sent two recoilless guns, so I remained with two recoilless guns.

I took a half-track and began running in the direction of Israel Meir (the company CO). After 200 meters, I sank into the sand, and then I saw dozens of tanks and an infantry force coming towards us. I reported to the division what was happening… I divided the sector— the battalion deputy commander in charge of the right side and I took Israel Meir's sector.

The recoilless gun platoon and the tank platoon reported that within the Wadi they could see a column of tanks and trucks moving eastwards.

The Egyptian Third Brigade was moving steadily through the Wadi, where the 89th Mechanized Infantry Battalion was positioned to defend "Akbarosh" in the area connecting Wadi Mabuk and the "Izavon" road.

Maj. Itai Margalit "H" Company commander reports that

There were two companies in the battalion, an 81mm mortar platoon and three 20mm guns. We spread out two companies along the front in a southwesterly direction, one on the hills east of "Izavon" and the other west of the road.

Yonatan Geva's (commander of 401st Brigade, Observation team) passed us as they retreated from their positions and told us: "You'd better escape—the Egyptians are coming." Our observation post reported that it saw a dust cloud in the—and the battalion commander instructed us to direct bazookas in that direction. When the Egyptian tanks arrived about a km away from our positions, "C" Company from the 46th Battalion arrived from the north. The Egyptians assumed positions from the Wadi northwards and fired about 20 to 30 tank rounds at us. We saw Yaron's platoon cross the Wadi about 800 meters from the Egyptian tanks.

At around 1030 we saw Phantoms aircraft dropping cluster bombs—later we saw that they had destroyed an Egyptian D30 artillery battery. The Egyptian retreat began after the air attack.

Lt. Colonel David Shoval, the 46th Battalion commander, described the following:

When I arrived, I saw about 30 to 50 tanks moving through the Wadi. Yaron was on the left, and I directed Ira to stand to the right, west of me. Yaron fired on the nearest tanks. When I arrived, one or two had already hit and had begun burning. When Ira joined us, the two companies were firing. I divided the sector between them—I also fired about 14 shells from my tank.

We were in ideal firing positions; they seemed to be in a range of 1000 to 3000 m from us spreading out through the Wadi. They drove like idiots, spread out across the Wadi.

This firing continued for an hour or two. The whole time they did not fire on us. By the end, I saw 30 burning targets.

Lt. Colonel Doron Rubin the 202nd Battalion commander reported that

The 81 mm mortar platoon fired at the tail of the Tank Brigade and on 339th Battalion forces that were near the battalion enclosure.

We all worked with the same radio network, so there were problems when many people were spiking at the same time, but everyone was in the picture, and the network did not jam. I reported to the division, telling them that I needed ammunition for the tanks, the 81 mm and the 120 mm mortars.

I received artillery support from the division.

At about 1000 hours, that phase of the battle ended. The division instructed the 202nd Battalion's tank platoon to return to the "Naveh"-"Yore" Junction.

The 89th Battalion was ordered to search the Wadi and destroy whatever remained of the Egyptian force; the 46th Battalion was ordered to cross the Wadi southwards towards the "Ze'eva" Road and search the area; and the 202nd Battalion was ordered to search the area in his vicinity.

The Egyptian 339th Battalion

The Egyptian 339th Battalion left the Ayun Musa area on its way to

Wadi Mabuk with the mission of moving along the 3rd Brigade's right (southern) flank. The "Naveh"-"Yore" Junction was not targeted by the Egyptian attack.

However, the force stationed there interfered with their advance plan and the 339th Battalion, which included three B.T.R. companies and a tank company (from the Third Brigade), began shooting on the 202nd Battalion's tanks that were firing on it from their positions. However, it did not attempt to storm and occupy the junction.

The Egyptian 339th Battalion commander lost control and direction; its forces split up, some of them, with the CO, continued towards the "Ze'eva" and "Hannibal" axis, where the 202nd Battalion, during that afternoon and the next day, finished the job, destroying the remains of the battalion and taking the CO prisoner.

1200-1700 hours–The Egyptian Retreat and the Area is Clear

The Division attention began shifting to the Gidi Road (the 164th Brigade) where a brigade-level attack was developed by the Egyptian 11th Mechanized Infantry Brigade that was reinforced by a tank battalion from the 25th Brigade.

The Fire Support Effort

Immediately upon the opening of the battle, the division mounted a fire effort with all the means at its disposal: mortars, guns, and air support.

As early as 0700 hours, the division requested air support; the Air Force jets began their attack 45 minutes later. The 202nd Battalion did not have radio equipment for air-land communication, and the attack method was by the division establishing a line that was west of it, where the air force conducted free attacks.

A key participant in the battle was the 209th Artillery Regiment commanded by Col. Ya'akov Erez, at the division's F.C.T., including the air support liaison officer who was a member of the Division F.C.T.

Col. Ya'akov Erez, the Artillery Regiment commander, thus describes the battle: "Most of the work was done by the air force. When the jets arrived, they requested that we not direct them . . . that we should let them work. They worked there, attacking freely for a long time and caused considerable losses." The air force attacks had a significant effect on the Egyptian force, regarding both morale and physical damage incurred on the advancing force.

The Battle End

What started unexpectedly and with a high potential for chaos ended with a great victory. The Egyptian Third Armored Brigade and the 339th Mechanized Infantry Battalion signed-out from the Egyptian forces.

A number of factors led to this success; first and foremost, was initiative and professionalism at all levels: in the battalions, the 202nd, the 46th, and the 89th; the combat intelligence system deployed by the 401st Brigade; the logistic support delivering ammunition to the battling units during the fighting; and the Division F.C.T. that managed and coordinated the battle.

The Egyptians also contributed much to this victory by their unimaginative and unprofessional management, on the systemic, tactical, and techno-tactical levels. The havoc wrought on the Egyptian commanders, especially the Brigade CO, knocked the brigade off balance and, in effect, halted the entire assault. From that phase on, the Egyptians were in "every man for himself" mode. Attempts of the chief of staff and the brigade's second echelon to come to their aid did not come to fruition.

Whereas on the Israeli side, the fighting was characterized by initiative and steadiness of purpose, on the Egyptian side, the fighting came to a standstill, and the commanders searched for solutions in the higher echelons. Still, until these materialized, they had no choice but to retreat.

The Egyptian 339th Battalion, which was surprised by the 202nd Battalion's presence at the "Naveh"-"Yore" Junction, fell apart before the battle got underway. Its forces scattered in all directions, some of them remaining to return fire at the junction and others moving with the battalion commander towards the "Ze'eva" Road southeast of the "Yore" Road and finally surrendering or being decimated almost without resistance.

Command and Control

Two deviations from standard battle procedures occurred during this battle. The first involved divisional battle management.

This battle comprised a tank battalion, a Mechanized Infantry battalion, a paratroopers battalion, an artillery regiment, brigade-level combat intelligence, brigade-level logistics, and air support—all of which were managed by the Division's F.C.T., while the Division commander was waiting for the Egyptians' main attack in a different sector (where the Egyptian 11th Brigade did indeed attack towards noon).

The battle was in an area of ten square kilometers against a reinforced Egyptian armored brigade, over two separate battle zones, the "Naveh"-"Yore" Junction area and the "Akavish"-"Izavon 62" area and Wadi Mabuk, where the 46th Battalion and the 89th Battalion were active.

Air support was conducted with no communication with the forces on the ground.

The fighting took place in a part of the Division control without a unique commanding network, since the divisional operational network was dedicated to the preparations for the expected Egyptian's main attack. The battle was managed by entry into the units' networks, under the direction, coordination, and control of the Division F.C.T.

The various areas of operations were determined in such a way that each force had its own sector, independent of other forces' activities, while its commander had the authority and freedom of action with a minimum of outside intervention.

All commanders possessed a very high level of professionalism and excellence.

The forces were instructed by the F.C.T. to enter the communications networks of adjacent forces, and, in certain cases, the Divisional team itself utilized company-level networks in order to coordinate movements and fire when forces from different bodies were active next to one another.

Supplying ammunition was done by the 401st Brigade; in the absence of its Division Main HQ and Division logistics officer, coordination was successfully performed by the F.C.T.

The second exception was the 202nd Battalion's battle management.

The battle created a number of command and control problems for the 202nd Battalion's CO. The effective forces at his disposal were two tank platoons, a recoilless gun platoon, a 120 mm mortar battery, and an 81 mm mortar platoon. They were all concentrated at the "Naveh"-"Yore" Junction.

The Egyptians attacked on two axes: the Third Brigade in a range of eight-to-ten km north of the battalion and the 339th Battalion in the area of the junction itself and, afterward, southeast of it.

The battalion commander was forced to split the forces beyond his ability to control them physically. He sent a tank platoon and a pair of recoilless guns towards the Third Brigade that was advancing through the Wadi and remained to fight at the junction with a tank platoon, two recoilless guns, and a mortar platoon and battery.

The tank platoon remained at the junction, could not maneuver since it defended the junction, and blocked the 339th Battalion's progress. The paratroopers' companies were useless in that battle. Although they were located in the area of the junction, they did not use their weapons throughout the battle.

The battalion commander managed the sporadic fighting as individual management, with personal bravery and endless initiative. He attempted to be everywhere at once.

In an inquiry after the war, the 202nd Battalion CO, Doron Rubin stated: "The battle was not managed at the battalion level, but everyone did what he thought was right, and it seems to me today as the best thing that happened in that war from every point of view.

The Battle in Numbers

I.D.F. forces:

- 202nd Paratroopers Battalion (2 Paratroopers Companies, 1 Support Company, 7 tanks)
- 89th Mechanized Battalion (2 Mech. companies)
- 46th Tank Battalion (2 Tank Companies, 29 tanks)
- 209th Artillery Regiment (2 155mm SP Btn, 175mm SP Btn, 120mm SP Mortar Battery)
- Air Force (61 sorties)

Egyptians forces:

- 3rd Armored Brigade (3 Tank Battalions, Mechanized Battalion, D30 [122mm] Artillery Battalion)

Casualties:

I.D.F.: 6 killed in action

3rd Brigade: unknown number killed in action, Two Tank Battalions (60 tanks), Mechanized Battalion (30 vehicles), D30 Artillery Battalion (3 122mm Batteries)

Bibliography

209th Artillery Regiment—Debriefings.

252nd Armored Division—I.D.F. Operations Log Archive.

252nd Armored Division—War Summary Journal.

Abed el Munim, Wazel. *The Israeli-Arab Conflict: Memories.* Translated from Arabic.

Amiram, Ezov. *The Grand Attack* (internal document, not published). I.D.F. History Department.

Avraham, Ayalon. "The October 14th Battle Deleted From History, 1978." *Ma'arachot,* Volume 266.

Dany, Asher. *War Limited in Size.* [Dissertation] Haifa University, 2002.

Doron, Almog. "The October 14th Battle in Wadi Mabuk, 1978." *Maarachot*, Volume 266.

Elchanan, Oren. *The History of Yom Kippur War*, 2003.

El Shazly, Saad. *The Crossing of the Suez.* Hebrew translation, 1987.

Golan, Shimon. *Decision Making of Israeli High Command in Yom Kippur War.* 2013.

Gamall, Hamad. *The War in the Egyptian Front.* Translated from Arabic.

Part 5

Mission Command Over the Horizon

Since mission command is culture, it will be with us in the future as it was in the past. In the twenty-first century, the technology changes are dramatic to such a degree as to define a new era: the Information Age.

The Information Age revolution is spreading as an evolution, beginning where electronic and digital technology forms the basis then slowly penetrating to other areas. At the end of the process, we are (or will be) in another age, the Information Age.

This applies to war as well. Armies have always leaned on technology, and technological developments have influenced the development of military doctrine. These developments include use of the horse, the bow, gunpowder, the internal combustion engine, and now digitization.

In the fifth section, we take a look at the future. Although we do not know what will happen in the future, the coming era will develop based on man-made technologies, so it is necessary to shape the ways we utilize technology. This section intends to inspire thoughts.

1 Mission Command and Non-linear Warfare

Brig. General Gideon Avidor

Introduction

It is impossible to separate the mission command approach from general military doctrine. The concept of mission command has been adopted, at least in part, by most armies, but consensus has not yet been reached regarding how to apply it.

In a document, David Alberts (2003) and others analyzed warfare in a net-centric environment and suggested possibilities and topics for implementing the internet in battle. Although issued almost two decades ago, this document is still relevant today. Much has been accomplished since then and greater experience has been acquired, but then, as now, we are only at the beginning.

Sometime in the future, our field of endeavor—in war and in general—will be administered by virtual systems. In a considerable part of our economic and social lives, this is already the case. Since war and military activity are among the most complex developed by man, however, more time and effort is necessary in order to realize the potential of technology for this field. This chapter will discuss this topic in specific relation to mission command in a net-centric environment.

The chapter will present the projected impact of developments in information technology on military theory and the need to shift from traditional linear to non-linear and single-dimensional to multi-dimensional thinking. This shift must primarily be based on capabilities rather than on hierarchical organizations focused on quantity and that are beset by uncertainty due to limited information.

The information era began in the second half of the twentieth century. Its development has been an ongoing process, which began with groundbreaking technologies that had a local impact and went on to influence every sphere of modern life. For the sake of discussion, it is possible to assume that this era will reach maturity in the second quarter of the present century.

Like all other areas of our lives, warfare has been affected by this process. In this chapter, we will primarily deal with one component of "revised" warfare, that is, command and control, and specifically with mission command. This is a topic widely discussed in every military organization, and it is not a new development; however, in an era of changing warfare, it has acquired new significance. In the first section of this chapter, we will present a possible scenario of warfare in the information era. There is no guarantee that these predictions will be realized, but they are still worthy of consideration.

Multi-dimensional Warfare

Warfare has and always will be a struggle between human beings functioning in an organizational framework with the aid of various systems in order to impose their power and exploit it to achieve their aims. In the present era, warfare is undergoing radical changes, most of them not on the battlefield itself, but in systems exerting their influence without first-hand involvement in the fighting.

It started with nations in arms during the Napoleonic Wars, that is, nations impacted by the war's outcomes when it swept through their territory. In the modern era—during World War II, for example—the impact of weapons and armies reached far beyond that of the Napoleonic Wars.

The new information era has opened up new dimensions. Most of them are non-military by nature, yet its technology has expanded the effects of war, sometimes with immediate impact and influence. This is mainly due to a physical process of urbanization and warfare within urban areas, reinforced by the information revolution that has caused the integration of bodies, resources, and systems of the virtual dimension into battle, including the international community exerting its influence through the media. The world was once organized according to nation states that waged war with one another. However, nowadays small, local wars are sometimes conducted between various types of organizations and nation states.

By its nature, warfare is not linear—as indicated by Clausewitz, who cited factors such as friction, luck, and others. It is waged by means of linear devices due to people's need to implement tools that they can understand and manipulate. Nevertheless, due to the limited human ability to determine how a war will develop, it is liable to end in non-linear chaos (generally for the losing side). This discrepancy creates an absurd situation where, although humans invented war, they do not have the ability to guarantee its outcomes.

Human beings have a limited understanding of reality and its repercussions. From the beginning of time, they have been aware of this fact and so have constantly searched for ways to improve their knowledge and understanding. Humanity has unceasingly changed, developed, and advanced in areas that are within the scope of its understanding, both by altering behavior and developing tools that improve its perceptions of reality.

Most human activities have been carried out in areas that are familiar, in a comfort zone—whether it be real or imagined. For thousands of years, people believed they were living in a world that was flat and that the sun revolved around the earth. This belief changed but not without difficulty, for example, at the cost of "heretics" who were burned at the stake.

Human beings function comfortably in a world they can measure, count, and weigh; that is, a linear, proportional world where if you give more, you receive more, and if you give less, you receive less. A world where everything is anticipated and orderly. Mathematics and engineering are based on these principles. Our lack of understanding of things from which we are distant enables us to conduct a full life, develop insights, and ignore what is "really" out there.

People have created a world for themselves in which their lives are conducted in a linear manner. What people fail to understand, they attribute to fate, luck, or supernatural powers. Warfare is no exception, it being a human invention that is generally conducted according to human linear thinking, even though it, again, develops in a non-linear manner. This gap generally is attributed to the "uncertainty" that arises when two armies confront one another on the battlefield, influenced by additional "participants" with various levels of involvement and interests. In such a situation, too many variables are present for human beings to immediately grasp and exploit them to their advantage. Thus, their failure is attributed to force majeure and uncertainty.

As war is a human invention, uncertainty is not an external influencing factor but the result of lack of information and lacunae in human capabilities. If we accept this situation without sanctifying it as force majeure and acknowledge that we prefer to function in a linear "comfort zone" that has no basis in reality, then we must attempt to break free of these limitations and find ways of functioning in a non-linear world.

Here technology can come to our aid; the information revolution deals exactly with this area and opens up new possibilities and tools to cope with uncertainty. However, although technology enables us to achieve better performance, it does not guarantee us victory. Technology is and has been developed to help human beings function more successfully and realize achievements that were impossible in the past. However, the human element is always present in warfare. The combination of technological capabilities and groundbreaking human thought can apparently raise us to new levels. The current meeting between human insights and new technological information systems points to new directions that can enable us to successfully command and control the non-linear phenomenon called warfare, namely, through mission command.

The world is non-linear and in constant flux, rendering human beings only partially capable of functioning within it. It appears that we will never be able to break into that world and exist fully within it, but we can create interfaces on the border between the linear and the non-linear and activate these interfaces in order to improve our understanding of and ability to function in an incomprehensible and uncontrollable environment.

Warfare

When the information available to commanders is limited, we focus on main efforts and secondary efforts based on our capability for linear thinking reinforced by human ingenuity and much luck. However, when the information available to us is much richer and readily available, we can leave more room for commanders to make on-the-spot choices based on their best judgment in real time and in a real environment.

Doing so calls for agile organization and flexible thinking based on successes that accumulate through the ad hoc introduction of reserves and concentrated efforts. The designated forces must be constructed autonomously for their mission; otherwise, it will be impossible to rapidly disseminate and combine efforts. Units are designated to achieve goals and prepared to efficiently combine information with other units for synergetic outcomes according to battle developments. Structure and organization must allow for flexibility. The centralized/decentralized battle combines tactical mission forces, reinforced by surrounding supporting elements.

The use of this swarming concept as a way of thinking means looking for weak points or opportunities, then bringing in one's forces and exploiting them. When the guiding principle is that the battle is fluid and takes place on a continuum, the power and wisdom of the commander will be grounded in his reserves. The leading force makes contact, breaks through or creates the basis from which the battle must develop, according to the strength and direction determined by the commander, at the time and place chosen for activating the determining force. However, this approach will be possible only when information is available about the enemy's battle environment and our own forces. This approach is best carried out if the mission command philosophy is well understood and practiced.

Warfare and the Information Environment–An Unbalanced Development

We study, analyze, and conduct warfare based on the information at our disposal. However, relying on information that we can absorb through our senses is precarious, as there is much more of it than we can comprehend or digest. In order to improve our achievements and view the general picture, human beings must overcome their limited abilities.

The gap between human understanding and the environment is called "uncertainty" in military operations and "force majeure" in other spheres. From the dawn of time, man was aware of this gap and never ceased his efforts to create tools to limit uncertainty, but it is still with us and apparently always will be.

There are two directions that may be taken to solve this problem: procedures and information. The first is an attempt to improve performance by means of teamwork, think tanks, and improved thinking processes. The second is the development of technological aids for collecting, processing, and disseminating information.

Until the middle of the twentieth century, such solutions were in the realm of disseminating linear information (such as print, radio, television, etc.) or were being developed in the sciences (such as physics, chemistry, and mathematics) that later provided practical applications.

The appearance of the computer constituted a giant step forward in information dissemination, followed by the personal computer and the internet that led to a new information age enabling mass sharing and application of information. One leading development direction was (and is) Artificial Intelligence, that is, teaching computers to "think" like human.

At the initial stages of the information era, these new capabilities were integrated into existing management systems. Today, far-reaching change is developing rapidly in every aspect of life. The information age has created a previously unknown situation in which we have at our disposal information of a scope, depth, and quality that greatly exceeds our ability to process it. In every technological revolution, time is required to assimilate and adjust to new developments and their application and to integrate them into individual and collective human value systems. It is far faster and easier to change a computer disc than it is to change human habits and culture. The gap between human conservatism and fear of change on the one hand and cognitive initiative and intellectual daring on the other may also be found in the military sphere.

Although military organizations are slowly implementing management and cognitive tools that have been developed and upgraded over the past two centuries, they lag decades behind what technology can actually accomplish. Whereas armies are already equipped with up-to-date and technologically advanced weapon systems, their battle doctrines, decision-making processes, and organizational procedures are at least twenty-to-thirty years behind the times, thus preventing the use of—let alone reality of—potentially improved capabilities.

One seemingly counterintuitive significant repercussion of the information era is a new perception of time; information can arrive at any location almost immediately. Time management has become a crucial factor in the efficient running of forces, resources, and command. This already holds true today and will develop much more in the future, when it will be possible to make decisions, determine timetables, and activate forces more quickly due to immediately available data and feedback in the field (as well as people being free of dependence on hierarchy). All is available to all at any time and space. However, all of this depends in military terms on revolutionizing organization, cognitive processes, command procedures, and officer training. All of which are the most difficult to accomplish.

Mission command, which was first formulated at the time of Frederick the Great of Prussia in a successful local battle, has become a central approach to warfare management in the information era. Fundamentally, it affords agility to the campaign commander by removing bureaucratic restrictions to the task for which they are responsible. Technological advances have reinforced this approach, making it possible to distribute and absorb knowledge that in the past was the sole province of the higher echelons and which was not easily made available to commanders at the front lines, thus limiting their ability to make educated decisions. Here, too, as in all decision-making processes, theoretical and organizational adaptations are likely to have a positive impact on much more than information only, as will be discussed below.

Traditionally, information has been (and still is) organized according to a command or operational hierarchy. This takes the form of a linear telescope that can cause delays, bureaucracy, and excessive control. In the information era, this telescope disappears and information flows on a horizontal plane, making it available to all. This change accords with the concept of the mission command system in that the higher ranks must slacken control, support subordinates, and trust them to carry out their assignments as well as possible.

Information technology is fundamentally electronic/digital, but the information era involves cognitive, revolutionary ways of thinking. Technology enables rapid treatment of information in every lifestyle by means of collection, processing, and dissemination. These developments are unstoppable, as they are not unique to a specific sector but have permeated every sphere of human life. In order to exploit these developments in the military sector, it is necessary to rise to a higher cognitive level, as the growth curve depends on thoughts, not on equipment.

Repercussions of Organizational Culture on Warfare

One of the most difficult problems in warfare was, and still is, a commander's ability to predict battle development and activate their forces as effectively as possible to achieve their goals. A number of central factors influence this:

- Insufficient knowledge
- Too much raw information
- Limited ability to isolate relevant information from an overload of sources
- Addiction to heavy computer processes at the staff level
- Poor training and vehicles, such as software or procedures, for efficient, real-time handling of accumulated data
- Poor awareness of situational developments due to a partial or incomplete grasp of on-the-ground conditions
- Lack of understanding of enemy activity
- The influence of unexpected or uncontrollable environmental conditions, including weather, public opinion, NGOs, uninvolved civil populations, collateral damage considerations, etc.

It is common practice to place all of the above in the "decision making" category under the term "uncertainty" as they limit the commander's ability to implement their forces most efficiently in order to achieve their objectives. This uncertainty is often presented as an excuse for operational failure. Typical solutions for uncertainty range from concentrating on ourselves and ignoring our surroundings to using "a lot," meaning "activating large numbers." There are countless examples of this, one being the ineffectual deployment of Israeli artillery and airpower in the Second Lebanon War of 2006.

It might be "comfortable" to blame failure on uncertainty, but this excuse is made less relevant in the information era. With innovative approaches and the help of new technologies, a different scenario might emerge where some of those problems might either diminish or disappear altogether. With the help of technologies for collecting, processing, analyzing, and disseminating information, it will be possible to narrow the gap of uncertainty in knowledge and to do so at the relevant time. The information era points to the need for solutions that will reveal problems or issues that were previously hidden from our view. In fact, when one deals with the general in order to arrive at the particular, one is constantly chasing after one's tail. In the information era, we can build a more complete picture from bottom-up details merging with top-down data. The changes in information's time and space—that is, when and where information is available—leads to immediate reaction capabilities in closed circles.

After all is said and done, there will always be uncertainty, but it will stem not from lack of information but from the fact that we are facing an enemy who is as determined as we are. Despite our ability to influence the enemy's decisions, we cannot fathom what he thinks or what he will do next. The center of gravity, therefore, is placed on the commander's shoulders more than ever. Quantities and equipment are not the decisive factors —as the last twenty years of fighting between states and non-state rivals have shown time after time.

We need to step forward into a new era, no longer involving upgrading existing instruments and systems, but instead making the transition to a very different future. In order to make this transition, a number of difficulties must be overcome. This chapter will now concentrate on the command and control element in battle conducting and management.

Information Relevance

This is not a quantitative problem but a matter of effective organization. The heart of the question lies in defining the information that is vital to the mission: What do we need to know and when? This decision lies with the commander when they visualize how the battle will develop. Sufficient resources exist today for selecting, processing, and disseminating information to support activities as required.

It is obvious in retrospect that, before every armed conflict in which we are caught unawares, there were plenty of signals that we failed to interpret. The Yom Kippur War is a blatant example of this failure, but it is by no means unique. Mission command is primarily a different concept of how to organize and operate command and control for battle planning and command. It leads every commander to fulfill their responsibilities, directs them to supply their sub-commanders with the vehicles to carry out their mission successfully, and provides the "big picture" to support their efforts and achievements.

Whether it is adopted practically or theoretically, mission command is implemented today as improvised problem solving; however, in the new information era, it should become the fundamental concept in command and control planning. To do so, we should consider the following.

Relevant Information at the Relevant Time

A surfeit of information, a rapid flow of events and a lack of tools for sorting, categorizing, and processing data at the appropriate time all create blockages and lacunae in formulating situation analysis suitable for battle management. The information revolution has produced innovative tools for managing, analyzing, and presenting data. It is necessary to characterize and introduce information systems tailored to the mission command concept for information processing and distribution based on the command and control method to support agile and rapidly changing missions and responsibilities.

Commander Training

Computerized information systems do not think or make decisions but function according to what is fed into them; the art of creative thinking is a strictly human domain. Even after artificial intelligence systems are developed, they will rely on algorithms that human beings have created. Because robotic decisions are purely technical, it will always be necessary for people to make decisions. Commander and officer training must be adapted to the rapid tempo of events and vast data accumulation. This means that training programs and technological developments must enable them to accomplish this successfully. In order to achieve this, simplicity is required in a unified and clear language, delegating responsibility, allowing margin of error, sufficient training and practice, and more. A starting point would be planning and organizing mission command systems to replace organizationally based ones.[1]

Mission Oriented Analyzing Abilities

Analyzing ability depends on the prior definition of questions demanding answers. When starting out from a point of uncertainty, analysis is performed in a generally determined format based on averages, and the results are calculated averages. We create these averages by setting end states and procedures according to what we think we are able to achieve. Such conceptions limit us in depth, length, and daring, due not to a basic lack of ability but a lack of operational and organizational information, understanding, and flexibility. In the new information environment, vistas open up in the big picture that were not previously visible. The horizon changes and there is more than one. In the new era, the panorama that spreads out before us has broadened considerably and has become more comprehensive and complex. It is multi-dimensional,

1   Organizationally based systems based on professional corps, whereas mission-based systems based on necessary capabilities, such as hand-to-hand combat, fire battles, etc.

with multi-participants and is rapidly and constantly changing. We are active players in it, no longer blind to the warfare environment. This holds true for all levels and creates new responsibilities, as we are free to make choices and focus less upon ourselves only.

Mission command can enhance that new situation and cancel the need for centralized control, as a hierarchical system cannot manipulate and successfully function in such an environment. Thus, the mission command concept should be constructed as a general system rather than a local solution. When we realize that we have tools far beyond our imagination that can enhance our powers, we will dare to think ahead and accomplish far more. According to the present model, we search for the "end state," that is, how to complete the mission and achieve its objectives. Through non-linear thinking, the mission command approach assumes we will approach our objectives through decision points. As each decision point is reached, new ones will ensue that will lead us to the next point, where the process will start anew and so forth until the end state is achieved. Information is our main asset and, at the same time, our main obstacle. The scope and quality of information needed will be limited to what is necessary for each step; the final mission is in the background, guiding us on our way. Thus, the necessary information that is gathered and processed will afford us freedom of decision and enable us to operate in a mission command framework.

Mission command narrows the uncertainty gap by developing non-linear networking concepts that operate in smaller, more manageable systems and function on all levels without losing control. On the battlefield, numerous factors are out of our control, including the enemy and the environment. Nevertheless, with a more "mission oriented" design, we can predict their behavior and exert our influence on them. The new information era provides us with much more information in a much shorter time but presents the danger of information overflow as well. We will never have absolute control, and we must direct and limit the amount of information we need; control of information is the vehicle of the entire operation's control.

The enemy is a hostile partner. The enemy's cognition functions under the same limitations as ours; thus, it is possible to integrate the human element into both planning and operations. Mission command and non-linear thinking place heavy emphasis on the cognitive element. Each point is a starting point; at each point, the enemy might face the unexpected, surprises or changes in directions and power. Because our fight is against human beings, we must consider this.

Warfare in the information environment provides opportunities to gain local superiority in information (intelligence), operations (surprise, temporal and spatial superiority, superiority in power ratios), and control (initiative, stratagem, exploiting opportunities, and overcoming crisis). This is especially relevant in asymmetrical warfare (against an enemy functioning in the framework of terror or guerrilla warfare). Such an enemy will perform numerous tactical, decentralized missions, while avoiding direct confrontation on the battlefield. It will employ technologically uncomplicated weapon systems and strike at times and places that exploit the element of surprise. Because it enables strong, decisive, concentrated action, mission command can cope far better with such asymmetrical conditions than can conventional military configurations.

If we improve our capabilities regarding the elements discussed above, the uncertainty gap will narrow to the point where we may enlist "force majeure"[2] to work in our favor. Information could be shared to advantage with non-military factors in the fighting arena—including non-combatants, economic bodies, and public awareness—so as to integrate them into a purposive, unified information system. This approach may be considered as the continual, integrated fourth (or fifth) dimension of information warfare.

2   As used here, "force majeure" applies to a given situation that is not subject to control at the time and place of the military operation. However it might be subject to change by various means.

The Role of the Commander

Mission command focuses on the commander and how he grasps the fighting development. Resources are not the main issue; their role is to offer the commander capabilities considerations for their decisions. The commander's field of expertise is in understanding the enemy, predicting how the battle will develop, and assessing enemy commanders' fighting spirit rather than in concentrating on details about his deployment. The mission command concept starts with the commander's decision-making, but its tactical organization operates, again, in a non-linear manner. Mission command is a holistic concept with two main functions: (1) in commanders fighting in their own way supported by the higher echelons' decision-making process and (2) supporting means allocation. Where does all this lead?

We understand that, when planning upcoming operations, we must act based on many short-term, coordinated actions. Of course, there are long-term goals, but there is no rigid pre-planning of how to achieve them. An old saying states, "Any plan is a basis for change," but here the changes are the plan.

An effective operation on an ongoing, multi-dimensional battlefield must include both long-term planning and a battle plan made up of numerous short-term actions. The objective of this approach is to arrive at each milestone with optimum conditions that will enable advancement to the next one. The plan is flexible enough to accommodate change so that various circumstances are created, disappear, or become more or less important according to developments. This approach demands close supervision of events and improved control systems that direct activity in real time and recommend the necessary adjustments.

This approach affords the commander in the field considerable freedom to decide, accompanied with relatively close support on the part of staff monitoring long-term developments. This approach influences not only field tactical command but also higher-level decision-making, staff work, and planning. It demands broad integration of various bodies based on the degree of their influence on how the operation develops, while requiring headquarters being capable of efficiently analyzing and processing data in the required period.

Because of this innovative approach to warfare, a new kind of work relationship must be created between superior and subordinate levels, one that is more akin to apprenticeship, accompaniment, and support than to authoritarian, proscriptive leadership. The upper ranks must create and sustain conditions under which subordinate officers will be able to sustain successfully a mission command framework.

In order for a commander to do so, they will need maximum support from the next highest rank regarding information and resources. The information era makes it possible to construct information networks focusing on different areas and distributed for various purposes independent of geographical location. A commander might be aided by information arriving from overseas for tactical battle purposes. The information era also enables maximum precision in aiming fire; therefore, it expands the fire support availability based on effective range rather than proximity, thus minimizing the need for a large auxiliary force following the fighting units.

Nowadays, the army must find solutions to a variety of challenges within the same mission, ranging from conventional battles to combating terrorism to stabilizing missions at the end of operations. Because we cannot establish a special army for every need, we need to organize forces and construct doctrine suited to flexible, modular decision-making and operational forces.

Staff and decision-making processes will need to accommodate such an innovative approach. Headquarters—in their decision-making, supervision, and control capacity—will need to adapt to such a form of operation and support the forces in carrying it out based on their specific fighting conditions and battle understanding. The staff needs to organize into subject-oriented teams to meet the concrete needs of specific missions—such as firepower, maneuvers, logistics, communications, and intelligence—in addition to integrative teams that will direct the operation by managing the current battle, planning the next one, or initiating special operations. In this form of headquarters, the staff's need "to support the commander" is still there, but its main activity is battle control. Thus, it takes on the character of a fighting element in its own right, allocating supporting means, "building" the next step, operating "out of nowhere" fire, intelligence and logistics support, and weapon systems supporting commanders at all levels in carrying out their missions. The headquarters is a crucial component in the networking combat arrangement. It operates on both hierarchical command channels and flattened operational support networks.

The battlefield is simultaneously conducted on two major planes that include countless interim battles and systems:

1. The physical battlefield includes geography, population, economics, the army—including its organization and equipment—the bureaucracy, and the control mechanism running those systems. In other words, this a kind of hard engineering casing with clear, defined directions of movement and behavior grounded in a set of rules with clear boundaries.

2. The information battlefield, including information centers (or centers of power[3]) created by and for the physical system but not necessarily rigidly attached to it. It operates as flexible networking connecting centers of powers based on interests. Its agility provides its ability to change constantly at a much faster rate than hierarchical connections can within the physical system.

Both systems are present on the battlefield, and the art of war involves employing a combination of the two according to command and control methodologies.

---

3   In this chapter, power means "influence." Influence might have many forms, in the physical, virtual, or cognitive dimension. Centers of powers in this paper relate to whatever has such an influence outside its own entity.

As long as we did not possess a wealth of information and the ability to apply it, decision-making was "made easy," and, in cases of uncertainty, the guiding principle was to send in the greatest forces possible. In the age of industrial warfare, wars were often won by quantity, whereas in the information era and with the appearance of irregular fighting concepts, quantity might constitute an obstacle. The information battlefield demands attitudes to time, space, and quantity that differ from the physical system. Although loosely connected to geography, it strongly impacts human cognition.

This type of battle is multi-dimensional and, again, is affected by non-linear influences. It consists of dynamic, changing forces linked by networks that are not subject to rigid, static measurement.

If we acknowledge that we are living within two parallel systems, questions arise regarding how to integrate them into a decisive physical-information interface. We might integrate them gradually according to our level of understanding, not as a revolution but as a gradual evolution. An example of this approach is the combination of electronic media, like television and social media, and political culture and public opinion. Net superstars have become trendsetters, and politicians curry their favor. If in the past, the electronic media served the purpose of informing the public and conveying the establishment's messages, this process now goes two ways, with a constant flow of information from the public to the politicians. "Physical" technology has caught up with cognitive development, and this process has just begun. As it involves interfacing between two different systems, this process is not a "natural phenomenon" but one created by and for human beings.

Today, we are exposed to a very different reality from that of the past, in which old explanations have become irrelevant. Our task is not to create tools that "tame" this new world to fit our old conceptions but to develop new insights, not to upgrade what exists but to take a bold step forward into the future. For hundreds of years we sought decisive wars meant to eliminate the enemy's fighting capabilities; in this information age, we might reach our goals by eliminating the enemy's will to fight—achieving that outcome calls for very different means and methods than those of the past.

However, if this process is not fully understood and is carried out ad hoc to meet a specific need or solve a certain problem, the result is a tangle of solutions that make it difficult to apply the new capabilities on a large scale. It becomes a case of "too many trees hiding the forest."

Today we are on the brink of improving our capabilities, but, since we cannot change human nature, we must equip ourselves with tools enabling us to create interfaces for exploiting what is beyond our comprehension. Again, at the "other side" of nonlinearity stands chaos. If we step into the nonlinear zone, then we must restrict our steps forward. Regarding military systems, command and control tools fall into this category, whereas dynamic, flexible command like mission command constitutes a higher level of traditional linear command precepts.

Since these potential developments exceed our ability to fully comprehend them, we must find ways of going forward and employing them to our benefit. Limited human capacities tend to create interfaces that, again, turn the virtual and non-linear into something that can be measured and controlled that we can "live with." Such interfaces mainly involve systems for processing and analyzing information and rendering it applicable to human purposes. As the human brain tends to translate concepts and non-linear outputs into the linear systems to which it is accustomed and according to which it can activate devices in the physical world, it requires an interface that creates a dialogue between non-linear and linear systems.

The correct manner of employing these tools would be to combine human control and flexibility without exaggerating in either direction, reducing bureaucratic regulation on one hand but avoiding total freedom and anarchy on the other. The mission command approach can create tools by which to maintain regulated hierarchical command integrated with flat networking control.

In the information era, a tremendous burden is placed on the commander, as the constant flow of information might engulf him; the issue is not collection but processing. In a hierarchical command system, information streams through a central "plumbing" system that is prone to clogging. The mission command concept demands commanders at every level direct data in such a way as to achieve the required results. According to hierarchical concepts, information is directed by the higher level toward subordinates; according to mission command networking concepts, commanders at each level pull the needed information from the data power center. On the one hand, this places a heavy burden on the commander, who now needs to define what their information needs are and make sure they receive them; on the other hand, they can tailor the information to their needs and plans.

Adapting Language as a Condition for Understanding and Control

As was always the case in the development of human culture, a language must be developed that is capable of describing reality and allowing the development of more appropriate behavior. Our world of linear concepts is two-dimensional and insufficient. It must be expanded to describe a virtual world that cannot be mathematically measured or weighed and is fundamentally impermanent. It must be able to describe concepts that are not absolute and whose central axis involves processes, not results. There are no "black and white" categories because anything is possible in the proper conditions; what is true today might be very different tomorrow. Even without our being able to describe such a situation in a way that we can comprehend, we can still use it to our advantage. Concepts can aid the operational organization in developing and exploiting situations and making alterations according to developments, thus exerting influence but not necessarily "conquering the mountain" to reach the final goal. Language can channel thought in new, creative directions.

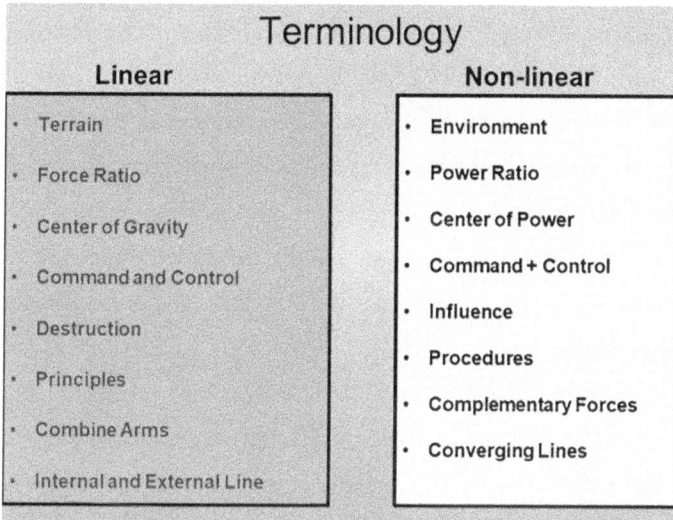

Terminology

Linear	Non-linear
• Terrain	• Environment
• Force Ratio	• Power Ratio
• Center of Gravity	• Center of Power
• Command and Control	• Command + Control
• Destruction	• Influence
• Principles	• Procedures
• Combine Arms	• Complementary Forces
• Internal and External Line	• Converging Lines

Figure 9: Changing Terminology

Recognizing the Temporary as Permanent and Preparing for What is Next

If, on one hand, we know that the outcome will be different than we expect and, on the other, we know that our performance can influence outcomes, then we must not exert all our initial efforts in achieving the intended result. Rather, we must invest sufficient energy in gaining superiority, while constantly examining what is happening around us, thereby retaining sufficient reserves in order to gain the initiative and progress toward the next higher goal that was determined in advance.

Of course unusual circumstances are liable to occur, in which over-investment is likely to avoid "distracting" developments—such circumstances, though, mainly occur where one side has a clear or absolute advantage. In a non-linear perspective, there will always be the next phase, and it will always be a critical and decisive one from which our further plans are derived and achieved, leading us on to our goal.

All our plans are subject to an expiration date, especially since we are confronting a rational foe; this being so, we must understand that absolute superiority is transitory and subject to quantitative and qualitative limits, as a wise enemy will always find ways of overcoming them. For example, the superiority of regular armies with conventional fighting power has caused terrorist organizations and less powerful nations to transition to semi-regular urban warfare, while recruiting forces and resources that do not rely on military might. In this manner, they have neutralized the ability of regular armies to achieve absolute victory; some examples of this effect include Vietnam, Afghanistan, Lebanon (with Hezbollah), and Iraq.

The argument between "sufficient" power and "absolute" power in warfare (that is, influence versus annihilation) is a long-lived one. On the eve of World War I, Sir Julian Corbett claimed that on the ocean "the first goal of the fleet is to ensure that the sea routes remain open in the most economical manner." Alternatively, Sir John Fisher, the first Admiral of the British fleet, was of the opinion that "the fleet's first priority is to find the enemy fleet and destroy it" (Falls, 1960).

On land, Clausewitz and other military theorists claimed that the goal of war is destruction, so it is necessary to concentrate and destroy the major enemy force, after which a situation will be created in which victory and other goals may be realized. Clausewitz wrote, "To achieve victory we must mass our forces at the hub of all power and movement, at the enemy's 'Center of Gravity'; thus will we reach our objectives."[4]

Continuous Warfare Having No Clear Conclusion or Single Center of Gravity

In military planning, two concepts serve as the basis for detailed operational preparation: "end result" and "center of gravity." They create a framework for planning and running operations, based on the idea

---

4   Clausewitz, Book IV, Chapter 11, p. 258.

that it is possible to determine "anchors" in battle development whose attainment will guarantee operational successes. This idea is based on a planner's point of view.

In a situation with no clear finish line, there will be no center of gravity in the accepted sense. Rather, there will be centers of power with varying degrees of importance according to the level of fighting that develops. We are witness to decision-making centers (on the tactical and operational level) and centers of power (on the operation and strategic level). The former primarily serve for planning and managing the battle, whereas the latter serve for planning and managing the entire operation as they have critical influence on the campaign. Today, when planning and activity are carried out with linear vehicles unsuited to a non-linear reality, a conflict arises between the organizational system and reality, a conflict described as "uncertainty" or "battle fog."

Rather than determining instruments that are comfortable to use and adapting warfare to them, instruments must be constructed that conform to the reality of the battlefield that can then form the basis for our thoughts and actions. This approach resembles driving either from one point to another on a well-paved and controlled highway or taking a cross-country shortcut. One is familiar and relatively safe with no surprises; the other leads from point to point, allowing changes in direction and wider possibilities.

Heavy armies overloaded with headquarters and complex, hierarchical decision-making systems do not support a freer cognitive style, as it goes against "proper order." Heavy organizations need stability and definite procedures and methods subordinated to command and operative principles with which they are comfortable. The encounter between Rommel and Montgomery in the World War II Western Desert campaign (1942) clearly exemplified this.

The linear approach works well when both sides are conducting symmetric warfare but not when one side (or more) breaks all the rules, driving the regular army to complain about asymmetry—as occurred when the I.D.F. confronted Hezbollah in the 2006 Second Lebanon War. Stubborn armies (and most of them fit that description) attempt to find ways of squaring the circle, of how to introduce asymmetry into the playing field, but when the means of doing this are unclear, unsuitable, or based on insufficient knowledge and it doesn't work, we fall back on quantity. If it doesn't work, send in more.

In consecutive fighting, there is no end state. Every situation is a beginning that flows in a number of directions, a variety of missions with various strengths and no center of gravity. Rather, accumulated situations flow from many channels, while the center of gravity for decision-making is virtually and temporarily created according to the nature of the mission. Commanders, in effect, deal with a battlefield impacted by forces, contingencies, local situation, and other environmental conditions. The flowing plan is best constructed for change from the outset, containing many decision-making points in which it is possible to design a different type of battle and construct a "new" task force from a variety of teams. In other words, mission command-style planning is tailored to ceaseless adaptation. Obviously, the tactical level cannot possess endless flexibility; thus, the appropriate mission command structure is based on firm organization at the unit level and maximum flexibility at the support and combat support level. Since, in the information era, we operate in the form of networking systems, we are not restricted to what is available on hand; support in all forms can arrive from a distance, as long as it is in range and linked with effective communication.

The Operational and Strategic Level

The national security arena in which we live differs fundamentally from what it was two decades ago. The clear boundaries between civilian and military (strategic-operation-tactical) have considerably blurred, and the distinction between lethal and non-lethal wars has become less clear than it was in the past. If in the past they appeared in succession, nowadays they are combined one within the other.

The involvement of distant contributory factors has increased, and the dependence on allies has become a more vital commodity on all levels than it was in the past. The influence of non-combatants, the media, and non-government bodies may be found at every turn and might drastically affect freedom of action.

The war does not end when the guns fall silent but carries on by other means. After achieving their war objectives, the army moves on to other types of activity. Wars conducted by such task forces as the coalitions that fought in Bosnia, Iraq, and Afghanistan emphasize the need to integrate stabilizing operations into their combat doctrine, that is, integration into the campaign after the fighting. A few weeks' fighting becomes a prolonged military effort that continues over many years, as was the case in Lebanon from 1982 to 2000, Iraq from 2003 to 2011, as well as in Afghanistan, Bosnia, Chechnya, and others. The I.D.F. has also adopted this concept into its doctrine. Operations found in the security zone with different degrees of severity and definitions range among crises, disputes, terror, guerrilla activity, limited war, and more. Battle doctrine demands an answer for all of these situations.

In most cases, it is the same army that needs to adapt to the new situation, so that army needs a doctrine it can adapt to this reality. An open framework is needed that is sufficiently flexible to accommodate non-linear fighting, thus, mission command and a doctrine based on net-centric warfare appears to be a good solution.

Overview

One way of examining the matter at hand is to determine a definition of war. Countless books have been written regarding this, with countless suggestions being made, ranging from quantity (size of force, amount of damage, and number of casualties, etc.) to objectives (political, military, national, etc.). In many cases in the past, wars were defined and evaluated according to the amount of devastation and number of casualties incurred.

Today these clear boundaries have become blurred, while a good number of military and political objectives are achieved by the combined forces on the "ground" as power demonstrations (in many cases by proxies) and threats of using lethal or non-lethal force. The repercussions of this unclear situation are inevitable, on both a theoretical and practical level. The distinction is not dichotomous: the infiltration of influential non-lethal forces onto the battlefield has not canceled out the use of violence but has changed its nature by adding new kinds of power tools. The non-lethal weapon system arsenal has expanded dramatically and has created an additional, that is, virtual, warfare environment that influences the battle sphere in similar ways than other developments, like air and space dimensions, have done in the past.

Nevertheless, war has remained the imposition of one side's will on another by use of force, although many elements have been added and the foe has changed its shape and function. The ratio between lethal and non-lethal has changed, so that, when destruction is dominant, the fighting arena is minimized to the site of destruction; when the non-lethal is dominant, the fighting arena broadens and deepens so as to make the lethal element only one component with a gradually decreasing relative value. Armies that were accustomed to reach decisive victories by destruction must now accommodate to a very different reality.

It is possible to discern this difference in the relationship between civilian and military components in wars of recent years. If in the past, the military component constituted 80 percent of the war effort and the civilian component finished the job, today the military component constitutes about 40 percent of the war, which is mainly creating conditions enabling the civilian component to set out on the long road toward achieving the war's objectives. One may discern this change by observing a war from beginning to end; the decisive battle is relatively brief and employs lethal resources in a concentrated geographical effort and relatively concentrated forces, whereas the final battle for achieving objectives is prolonged, decentralized, and ongoing, integrating civilian elements and selective resources. Together, in varying proportions, they end the war, but the army is still chiefly responsible for creating the conditions for "ending the war."

The rise in value of the non-lethal component has changed the world order. No longer can one hierarchical framework acting under one commander achieve success on its own. There are no direct connections between results and centers of gravity no longer exist; rather, everything is in flux, with continual fighting among constantly shifting power centers. The lethal effort is limited and finite, whereas the non-lethal effort is characterized by prolonged interim situations manipulating and supporting lethal outcomes.

Fighting by Proxy

With the realization that conquest or destruction in their various forms are the result of many actions and are not necessarily a culmination point, comes the concomitant understanding that it is possible to achieve military objectives by other means. When it is crucial to limit casualties and collateral damage and avoid political complications and enraged public opinion, it is preferable to act through a proxy. Doing so enables continued pressure on the enemy and ongoing fighting without directly investing forces and resources, even without holding nation states responsible.

This strategy is certainly not new, but, in the past, it was mainly done on a hidden strategic level. Today, ongoing fighting takes place on both the tactical and operational levels. At various times, Russia, China, Israel, Iran, and many other countries have implemented such strategies. Here, also, mission command is appropriate; the "proxy" is assigned a mission and performs it to the best of their ability, while the objectives are defined from a distance by the initiator. One form of proxy derives from information era weapon systems to the point where it may be considered a weapon system itself. Information management becomes a basis for a new form of warfare: the information war.

What is the Significance of This?

If fighting is consecutive and the means of achieving objectives are ongoing, then concepts, systems, the commander's behavior, and the staff headquarters must all be adapted to this reality. The entire theoretical and practical system must be prepared for fighting in perpetual motion, with changing forms and methods, which means that the command and control system must enable and support perpetual change. Constant fighting demands clear definitions; for example, the term "maneuvers," which was previously defined as a combination of movement and fire, now receives a new meaning. It now means a combination of all the elements influencing the objectives via achieving supremacy and control on the way to reaching them, including movement, fire, information, deception, shaking the enemy's defenses and the morale of its troops, interfering with reserves and fighting strategies and whatever else imaginable is dedicated to the same mission, regardless of its original chain of command. Ongoing maneuvering is multi-dimensional and takes place in a simultaneously physical and virtual environment, with varying degrees of one or the other dominating, but with their always working in harmony toward a unified mission under the auspices of single commander.

When results and centers of gravity are the cornerstones for planning and building a maneuvering force, the fighting plan will be coerced to conform to our conceptions of the battlefield while ignoring those of the other side. We will pay the consequences, as the Second Lebanon War of 2006 blatantly exemplifies. In the information era, we must organize forces and resources differently and develop planning and management processes suited to acting in a network-centric warfare environment combined with a constant fighting environment. This necessity leads us to a "no-man's land" between linear and non-linear warfare, and the more control over it we achieve, the better the results. The "no-man's land" is there all the time; we need to adjust to operating in it.

The way to gain control of this no-man's land is to construct systems and doctrines that overcome our cognitive limitations, to perceive, digest and act effectively in the necessary period and dimensions. Instead of operating with a limited number of large, heavy bodies, doing so will demand activating many small teams acting in synergy and harmony toward a single unified goal. This unity begins by organizing a chain of command and net-centric control and information flow.

According to the traditional linear concept, we have five-to-ten elements in a formation relying on a set of vertical channels for command, control, and support. We build all kinds of "supporting" secondary systems to bypass checkpoints and shortcuts. The new information capabilities with rapidly streaming data creates overload even before processing starts, resulting in traffic jams and clogged arteries. A way to open this situation is by splitting linear telescopic channels into specialized information centers supporting intermediate missions. The same idea applies to organizations supporting continuous fighting, with continuity and flexibility being achieved through the extensive use of modular building blocks, strong reserves, and maneuverability as the leading concept in all systems, not just by a powerful breaking force. Commanding and controlling that way of fighting can best be conducted by the mission command concept as the guiding force for building, training, and leading.

Penetrating the Non-Linear Battlefield

Network fighting with a non-linear approach relies on information age capabilities and constitutes a step forward in the art of war. The slogan "Every plan is a basis for change" serves commanders following a linear approach as an excuse for lack of planning or failure. In a non-linear approach, we prepare and plan for changes. We plan smaller steps, being ready to change direction regarding means and power; we organize a force with reserves to introduce according to a developing situation instead of preliminarily apportioning forces in a rigid manner. The mission command concept is established and built into the organization as the leading doctrine of activating forces, as it makes non-linear warfare possible.

2 The New Dimension in War– Virtual Warfare

Brig. General Gideon Avidor

The future discussed in this chapter is one in which the technological information revolution reaches maturity and its applications are available to all. This process is not purely technical. Tomorrow's wars will have to adapt themselves to acting in a different cultural, economic, and political environment. War has always been multi-dimensional, but in the past, it focused on direct confrontation with an enemy within a defined contact area. As we have seen, the Information Age fundamentally changes temporal and spatial concepts that were prevalent in the past.

The relative role of those involved in warfare has changed, the distant has closed, the influenced have become the influencers and a new balance has been reached. The Information Age is creating new situations, starting with a new virtual dimension (information warfare) and ending with tactical and operational technological capabilities of a weapons system in a wide range of functions.

The virtual dimension stands alone and includes two components. The first involves providing support and assistance to the forces via networking as net-centric warfare, including communications systems for elements involved in the combat space, such as weapon system intelligence, fire, logistics and the like, headquarters and commanders, and data processing and management. This component is an integral part of ground forces operations.

The second involves cyber systems and electronic warfare of various kinds. This warfare is directed against enemy systems and commanders with the aim of gaining superiority at operational and strategic levels. However, it is an independent, stand-alone battlefield, comparable to the air force and the navy. It may provide support of—and integration with— ground forces, whether for attack or defense. Missions, including patrols, security, tactical intelligence, deception or security, are not foreign to this dimension.

Warfare in the virtual dimension might take place without any physical fighting and might be non-destructive; for instance, Chinese combat doctrine preaches how the weak defeat the strong by means of information warfare (Thomas, 2003). It provides an additional dimension to war that may be integrated into battles on land, sea, and air, from the lowest tactical to the highest strategic level—all with the aim of achieving superiority. Nowadays, we are only at the dawn of these developments, so the virtual dimension is not yet perceived as having an independent existence. Rather, it is utilized by various specialized organizations or integrated into existing ones. In the future, continued technological developments and their integration on all levels will cause it to be recognized as a separate dimension, as were the air, army, and space programs.

Adding the virtual warfare dimension can sustain continuous fighting, free of human physical limitations like fatigue or anxiety. It has no geographical or topographical boundaries. At periods when direct contact by maneuvers or firing ceases, fighting continues in the virtual sphere anywhere it is required. This form of warfare is already being directed against national and economic infrastructures, as has been seen in Chinese and North Korean cyber-attacks on industry (Thomas, 2003), and this is just the beginning.

Multi-dimensional warfare that combines physical and virtual dimensions acting in harmony and synergy under one supreme commander creates multi-capabilities and demands continuous fighting. This, in turn, demands a suitable deployment of forces and resources, command and control. Deploying troops in a decentralized manner enables concentration or dispersion as needed, with accelerated advances or changes of direction limited only by the commander's will and the quality of the organization (and the enemy, of course). The link between the two dimensions is man, who dictates the quality of performance; this type of battle plan enables the commander to realize their highest abilities and those of his subordinates, with mission command as the most suitable format for doing so.

Typically, technological innovations precede human cognitive development, so their integration involves a lengthy process of trial and error, success and failure in tailoring these new capabilities to human needs. We are at the beginning of a period that started in the 1960s and has been rapidly developing ever since. We still have a long way to go, but clearly the past is long gone, and we must now look to the future.

A Hologram of the Fighting Arena

Technological advances create available data anywhere and at any time and for almost every need, both for decision making and operating technological systems. The opening conditions have changed, more participants having direct and indirect influence, some of them uninvolved in the actual fighting but having an impact on outcomes.

The fighting arena has also changed, combining or blending a physical, geographical environment with a military and civilian population, all enveloped by a virtual dimension supplying data, management systems, command networks, virtual power centers, public opinion, and the international and local communities. Command and control systems are now accessed through open and closed networks, thereby fundamentally altering and making redundant past hierarchical relations. In the past, these systems were based on information hierarchies. As Francis Bacon stated long ago (1597), "Information is power."

In a linear system, collected data was sent to the higher level for analysis and then sent on to "clients" at the higher level. Technology has created "looker-shooter" closed-circle systems based on equipment, comprising a small step in the right direction. The information era with net-centric warfare enables any user to pull necessary information without depending on a higher level's will or efficiency. The internet has shown us the way to realize this relative autonomy.

Of course, a surfeit of information can create "data fog," concealing what is important. This possibility demands the integration of information management systems whose function is to categorize and verify data according to topic and specific mission; this integration lies within the responsibility of the high command, not as a means of control but as a source of correct streaming of vital information.

According to the mission command approach, the information management system creates professional or operational centers of power adapted to the need of the mission or users and linked by network-centric systems. It assembles or dismantles these centers ad hoc according to the mission's needs. The net "boundaries" provide flood control or security, but they are dynamic and decentralized, supporting every fighting force, whether tactical or systemic.

Flattening the hierarchy of available information significantly changes the meaning of time and exploitation, while changes in the space dimension exert local influences from distant geographical spaces, thus broadening command areas and hierarchies to efficient network spaces for various near and remote power authorities. All of these changes create a new way of looking at strategic, operational, and tactical hierarchical arrangements.

Hierarchy consequently loses its sector significance in favor of content. The hierarchy is mission oriented, which is fixed; the order of battle, though, is agile. A blending of capabilities and direct communication occurs at the point where hierarchy is no longer necessary. Only two levels are needed, one apportioning missions and resources and the other performing them. Other intermediate "layers" slow the flow of operations and information. In this respect, less is better.

The mission command approach enables the construction of ad hoc fighting formations adapted to need and mission, which sometimes are lengthy and operational and at others short and tactical. Network systems enable this flexibility, with the physical organization of units and formations as either a limiting or motivating factor.

The Fighting Arena in the Information Age

The fighting arena is the outcome of military leaders' cognitive grasp of objectives and ambitions and their armies' force structure and technological capabilities. This chapter will not deal with the cognitive dimension but with an aspect of the technological one, namely, command and control in the technological Information Age and their influence. In the past, the "virtual environment" existed both as a battle of the wits among generals and as a matter of morale and motivation.

Nowadays, the Information Age "moves continents" and brings the heart of the battle into the urban environment, which has existed for thousands of years in varying degrees of importance according to each historical period. In the broadest sense, the urban environment was always a decisive factor when it was physically involved in warfare. However, in the information era, cities have become targets due to their being critical information centers of power-wielding, strategic virtual (and physical) warfare.

Urbanization has made cities the significant part, while the periphery, having lost much of its importance, exists mainly to service the metropolis. The cities have become the cultural, economic, and political core of countries, housing as they do more than half the population worldwide. The information war is directed at a number of centers of power, all of which are located in cities. In most instances, the military may be found outside the cities, as it prefers to fight in open areas. From the strategic point of view, the army is no more than an obstacle protecting the center of power. This situation is not new, but in the information age, warfare can bypass the army on its way to centers of power. That ability calls for new forms of warfare, virtual battles being among them, but the physical "face-to-face" battle will need to adjust as well.

The virtual sphere has added to the tactical and operational fighting arena a range of virtual activities as integral parts of physical warfare. It has added new types of objectives to the struggle for physical and virtual supremacy, while expanding the fighting arena and its activities range. The battlefield has always been subject to "external" influences, but, in the Information Age, external forces are no longer at a distance but are adjacent to, or even penetrate, the battlefield and directly impact possible actions and their consequences—cyber and electro-optical warfare being prime examples of this. Thus, ground fighting has not only become even more complex and concentrated than in the past but also greater in volume, width, depth, and height. Weapon systems have rendered complex armed struggles even more complex. This demands an attempt be made to simplify it, and one way to do so is to improve command and control systems as part of the new information age.

In our times, cities have become too big to swallow, unless one aims at flattening a city altogether—such as Grozny, Aleppo, and Mosul, to name a few. Fighting in cities leaves two open spaces: the air and information dimensions. Both can create superiority, but in order to achieve control over a city, one needs "boots on the ground." This need calls for multi-dimensional net-centric warfare, something that needs to be developed afresh. Important components of these new developments will be command and control methods and systems.

A situation is developing in which new thinking is necessary, no longer adapting past successes to present conditions but instead being open to a future where the unexpected and unknown are the central focus and ways of behavior will need to adapt to future capabilities. Mission command can lead to a new era characterized by effective multi-dimensional, mission-focused warfare free of hierarchical restrictions and bureaucracy.

Past, Present, and Future

Command and control have been an integral part of warfare ever since struggles involved two or more fighters. Armies have dealt with these elements from time immemorial up to the present day. In this chapter, we will now attempt to predict a future, while making a number of assumptions.

Dependence on the past can be relevant for processes moving forward to the future, but, at a certain point when a significant change occurs, a large proportion of experience can act as a millstone around one's neck. While considering processes, it is occasionally important to reconsider fundamental principles that form the basis of past solutions. In the information era, technological innovations generate cognitive and physical capabilities that can only be fully exploited when they are freed from the bonds of the past and form the basis for new structures.

For the sake of the present argument, we will assume the following:

- There are enough indications of how the Information Age will influence operational management: the internet and social media, global systems, and their like indicate these directions.
- The virtual arena will be governed by its own dynamic from which systems and resources will be activated to integrate or damage infrastructures and mobile systems; mobile smartphones, television, satellite communications, cyber systems, and their like are constantly improving.
- The integration of information-based innovations into technological structures will continue; robotics, sensors, and command and control systems are already at various stages of use and development.
- Technological systems have limited capacities, as they perform only what they are programmed to do. Meanwhile, they are incapable of thinking independently, drawing conclusions, and improvising. In order to derive the maximum benefit from them,

we must combine systems with different capabilities, and some of their capabilities must be integrated with other technical systems, whether for operating weapon systems or planning strategies.

Organization Adapted to Capabilities

The structure and organization of armies today are based on experience adapted to predicted threats and technological abilities. Over the past years, various slogans and solutions for increased functioning efficiency have gained and lost popularity, including "air-land battles," "full-spectrum operations," "a revolution in military affairs," and "multi-dimensional warfare"—each era having its own pet phrases. In the Information Age, military employment needs to be organized on a mission command foundation due to its dimensions, and it should be activated by net-centric concepts. These will generate combined multi-dimensional operational capabilities exploiting a combination of information-based resources, tailored to the needs of a mission.

Modularity and versatility will be the guiding principles at every level, and the combination of these building blocks will make up combat formations supporting mission command warfare. The doctrine will be formulated so as to enable agility in concentrating effort; economic employment of means and forces by net-centric warfare; and concentrated task forces on the unit and formation level based on exploiting successes, initiative, flexibility, and lethal power. Command and control requires building an information system, flattening hierarchies, and adhering to mission command as a leading concept.

The Influence of the Information Revolution on Warfare

Due to the information revolution, local and international weak points have become widely accessible, enabling quasi-military organizations to exert their influence on the general populace and affect all areas of private and public life, the economy, and security systems. Thus, a non-lethal public warfare sector has developed parallel to the lethal battlefield, and war is no longer limited to physical battles.

The goals of such a war are not limited to the struggle for material resources or a country's identity. Instead, they expand to include such social issues as imposing religion, culture, or ideology; the idea is to attack from within, utilizing local forces. A situation has arisen in which the gap between military victory on the battlefield and achieving war objectives is constantly widening.

As the information revolution gained ground toward the end of the last century, this type of warfare has gradually become prevalent. Amazing technological advances have changed the balance of power within and between countries and rivals. New forces have emerged, old ones have been downgraded, and the global village has made its mark on all areas of our lives. Nevertheless, armies and battle doctrines have remained among the most conservative sectors of human society. Although they undergo development and become a focus of interest in wartime, in peacetime—from a national point of view—they recede into the background and lag behind other domains in technological development.

Technological Developments' Influence on Military Doctrine

When studying military history, we encounter countless examples of new technologies that influenced military doctrine. An example would be the Mongols taming the horse, inventing the saddle, the rein, and the short bow, and arriving at the gates of Vienna. Other inventions include the wheel that enabled the development of the chariot and gunpowder that resulted in the production of small arms and heavy artillery which expanded the battlefield and changed the balance of forces within it. The internal combustion engine also effected massive advances, including ships and submarines, trains and planes, tanks and guided missiles, thereby expanding the battlefield to far distances over continents and oceans. Every development met with a counter-development, and the competition between weapon system and counter-weapon system has always played a part in war. Although not yet mature, the fast-maturing information era battlefield will, in addition to the influences within the battlefield, expand war to every aspect of our lives—both military and non-military.

The first signs of these developments appeared when warfare shifted from symmetrical fighting between regular armies to asymmetrical fighting between regular armies and terrorist organizations. Vast armies have found themselves helpless opposite an invisible enemy, and wars that in the past would have ended with decisive military victory now remain unresolved, leaving armies to seek new, ever limited solutions that would "restore the situation to normal."

Volumes have been written to justify this situation, with "asymmetry" being offered as an excuse for failure. Actually, in any war, we attempt to create asymmetry in our favor, so what has changed? The term "hybrid war" has been suggested as characterizing change, but it explains nothing. What has happened is that the enemy, by quick adaptation to information era developments, has found ways to outwit us by avoiding the heavy battle and instead moving directly "over our heads" to civilian centers of power. The search for solutions has raised the classical linear concept of "the more the better." This is a mistake, though, as the enemy has prepared itself precisely to counteract this approach by acting according to non-linear concepts, aiming at virtual victories in public opinion and social support.

Developing New Equations

The gradual developments of the information era have initiated new forms of military thought and introduced a new dimension: virtual warfare. The physical dimension has always been present in war in the form of maneuvering and fire power. (Also always present has been counterfire.) There has always been competition between fire and counter-fire fighting for supremacy according to the concrete situation at hand, whether one is on the offense or defense. In the course of World War I, airpower added to the equation, and air space quickly became part of the battlefield. But it took about twenty-five years for air space to become a warfare dimension in its own right. The debate regarding supremacy was limited to the physical dimension; at the end of the twentieth century, the Information Age appeared, slowly removing this limitation.

The "asymmetry" argument leads one to conclude that a quantitative balance of power does not win small wars. In the struggle between equals, numbers can at time determine outcomes, but when the forces are asymmetrical, the perspective changes. Striking examples of such change would be the I.D.F.'s 2006 confrontations with Hezbollah in Lebanon and against Hamas in the Gaza Strip over the past decade, where Israel's overwhelming numerical advantage did not prevent both sides from claiming victory. This is non-linear warfare, where every gain creates a new beginning, creating a political warfare capable of realizing the same political objectives in service of which the battles started in the first place.

If we hark back to definitions from previous wars, like end states and centers of gravity, we will see that the first represents the desired outcome of a certain operation, and the second, the source of physical and spiritual power required to achieve it. No matter how these are defined in the literature, regarding Army A or Army B, they are no longer relevant. End states are no more than opening conditions for the next phase, and centers of gravity are divided up into a number of sub-centers that, when combined, can provide the necessary power to achieve missions. They are not necessarily geographically connected but are linked by networks acting in synergy so should more accurately be termed "dynamic power centers." The new environment is dynamic and in constant flux, while the balance of power and influence is ceaselessly shifting and changing direction, along with other components that have bearing on the situation.

When every conclusion becomes a new beginning, the balance of power equation might well shift at the opening of each new phase. For instance, if we completed Phase A by fire, Phase B may possibly be achieved by maneuver, as the situation demands. Thus, our activities must adapt their emphasis, resources, and forces according to real-time conditions on the battlefield. In general terms, there is nothing new in this, but changes in time, space, and information require a new look at the meaning and doctrine of how we should act in the new era. In such a situation, command, control, and battlefield management systems take on new importance. Rigid hierarchical systems would have difficulty functioning in such a flexible, rapidly changing, and multi-dimensional battlefield. What we need is a dynamic, flexible system that can accommodate modular, non-linear warfare and a doctrine and forces structure that provides us with supremacy over the enemy by our gaining the initiative. This need will have far-reaching implications both for officer training and for military doctrine.

According to this approach, mission command and control systems are constructed to realize the mission through a wide, open perspective of power build-up and flexible battle management. We now need staff work and decision-making processes suited to flexible, conscious, and rapid action. We must distinguish between hierarchical command centering on command and mission command concentrating on control and management over networks. The control and management systems need to provide the commander—that is, the hierarchal system—with a vehicle for executing mission command-style operations. Both forms have their own rules and processes and must be linked by interfacing.

Multi-dimensional Inter-system Maneuvering

"Maneuvering" is a doctrinal blanket term covering insights, force organization, and operations doctrine. Its central principle is that, through initiative, aggression, and stratagems—as opposed to exhaustive battles, firepower, and attrition—human wisdom may improve a force's effectiveness and physical. A striking historical example of such battle maneuvering would be the World War II German blitzkrieg as a reaction to the trench warfare of World War I.

In the information era, when networks replace the decisive field and the focus shifts from people to systems—especially when fighting takes place in complex urban areas—maneuvering requires a new meaning. Maneuvering and other means—whether lethal or non-lethal—must bring the war to the enemy, not only physically but also virtually. The enemy's centers of power, which are the maneuvering objectives, constitute a multi-dimensional space. They include the enemy's command, control, and management systems; public opinion and image; morale and motivation of fighters and civilians; and the involvement of the international community. Maneuvers are conducted multi-dimensionally as well—on the ground and from the air—in an integrated effort of physical and virtual power and movement (including irregular warfare, cyber, remote-control means, etc.).

Maneuvers are directed against systems, attacking them by many means from many directions in a non-linear attack. Some of them are conducted with traditional fire and movements but, as a whole, with much more.

The common picture of a linear battle is one force aligned opposite another in order to conduct an "organized" campaign. Multi-dimensional maneuvering breaks down this order. Fighting becomes fluid and agile, especially when directed against an enemy that is not a regular army. The fighting arena spans 360 degrees and encompasses the virtual sphere as well. This expansion demands new ways of thinking and the development of appropriate tools not meant to support linear troop movements. This development should be based on agile, but effective, systems and calls for commanders capable of functioning with multi-dimensional maneuvering on a fluid, fast-changing battlefield. All of these tactics can achieve amazing results, but only if appropriate command and control systems are put into place. Thus, mission command—which was exceptional in the past—appears to be the main direction of the future.

Continuous Fighting in a Non-linear Environment

In order to function in a non-linear environment, we must define it first. To this end, we begin with this list of principles that we will elaborate upon later.

- "Non-linear" means constant change having no end state, but rather a chain of opening conditions in a limited period. This is actually a flow of events that are artificially "frozen at a given point" (place or time) according to a need.
- In order to function in such an environment, timing and structure are adapted to the mission at hand.
- The degree of change and its changing direction or power is influenced by the power and nature of the participants in the operation and their opponents.
- Power of participation is a general, complex definition comprising numerous contributory factors that generate effective power.
- Effective power is defined according to what is deemed relevant and not necessarily by the power of the participants, since, in different circumstances and for particular needs, the same participant will have varying degrees of effective power.
- The more long-term designated an operation is, or the larger the number of participants, the more difficult it is to predict developments and plan accordingly.
- Influence is the major result of an operation and is judged by power (type and mean) and movement (location, timing, continuity) and may be measured or assessed by positive or negative values.
- When an operation is planned in an environment that is in perpetual motion, the battle plan must integrate these influences

and control their development in order to achieve the desired result.

- Functioning in a constantly changing environment is net-centrically supported; thus, it combines forces and elements with cross-hierarchical boundaries. Therefore, it is necessary to be able to map relevant strengths in the field of action, including nodes or junctions. Activators—any agency bringing about activation—and power are the factors influencing the process that are relevant for our purposes.

- By means of mapping, an activation system will be created that will enable command and control over processes and allow agile decision making that improves the chances of achieving desirable results.

- According to the non-linear approach, such a plan of action will comprise a succession of small, flexible activities directed toward a unified goal and capable of changing direction at all levels.

- The general master plan defines interim goals, initially directs mapping, and later supervises direction of effort.

- Specific plans are directed toward action and constructed as short, controlled processes that can be combined, divided, and are mutually supportive or performed in synergy according to the need.

- Such a plan includes physical components, including active forces and the required interfaced systems—like command and control—to achieve its goals.

- Such an approach demands organizing mission command net-centric warfare principles, both in forces structure, staff work, and command and control procedures. Its effectiveness depends on combined staff teams and integrated command and control systems providing relevant information to those requiring it, wherever they might be and when it is needed.

All of the above is mission oriented, so analysis and management must be based on actual and expected results, not on resource investment.

Decision Making in a Continual Fighting Environment

The involvement of an additional virtual warfare element expands and increases the boundaries and duration of armed conflicts. Combat might remain unchanged, but warfare as a whole will change. Information warfare has no geographical boundaries or human fatigue restrictions, and it can integrate at any fighting level with multiple targets and goals. By introducing information warfare as an integral part of warfare as a whole, we will break through the "restrictions" of time, space, and human capabilities, making war continual. Another major influence of the information era is that, once we enter the information dimension, our standard control measures are unsuitable. A new fighting dimension creates multiple new options that, when used by various actors, take fighting out of our control and render it non-linear.

According to non-linear theory, at the end of the non-linear spectrum lies chaos, so we must gain control if we want to avoid chaos and win the war. It thus becomes clear that, in order to gain the upper hand, we must adjust our decision-making doctrine and practice to suit a non-linear environment.

If we understand that our control of the non-linear environment is limited, we have two ways of dealing with it. One way is force-centered control as we acted before the Information Age. The other is to construct a system suitable for rapid, unexpected changes that allows leeway for immediate, effective adaptations, as is the case with the mission command concept discussed here.

Since we can only partially control non-linear developments, the concept we must follow is planning a non-linear campaign while acting in a linear format: we take small linear segments steps that enable changes according to the (non-linear) results of our linear actions, after which we plan the next (linear) move. This approach requires organization suitable to dynamic and constant changes, without losing superiority or control, and a suitable command and control doctrine.

Consequently, we will determine a plan with many limited goals, each reached by a force operating according to mission command. We will climb from one small goal to the next on the road to the mission's objective. We are capable of changing and adapting along the way because the campaign is flexible, so we can concentrate efforts or disperse them at will—on condition that we direct field commanders to carry out their mission alone according to the higher command's intentions.

The dimensions of the goal (volume, distance, and complexity) will determine the extent of the linear steps that can safely be completed. Good planning will always consider continuity, including the next step or possible changes in any course of action.

Such a planning concept is universal, so that every operation on any level will conduct its planning and forces dispersion based on the same concept, namely, mission command and mission-oriented flexible forces deployment.

Building an Integrated Multi-layered System that is both Centralized and Decentralized

What we require is a framework for planning and action that is directed at war objectives and enables maximum operational flexibility without losing its general direction. Regarding the operational system, we may find centralization in an operational framework linking a network of milestones that serve as road signs and net-centric resources accessible to end users requiring data and support. We may find decentralization in applying coordinated mission command objectives to and from the milestones.

Regarding staff work, unity of command provides centralization, with one commander running the operations. Decentralization occurs from distributing authority to sub-commanders, each responsible for their mission under the higher level. Thus, a complete system is created which is centralized from the outside and decentralized from within. Forces and staff headquarters must adapt to functioning efficiently and economically within this format.

Coping with a Dynamic Environment

We are gradually realizing that the two-dimensional linear world to which we have become accustomed and in which it is comfortable to live and in which we have formulated our battle doctrine is not the whole truth. Another reality exists, one that is difficult for us to grasp, no matter how hard we try. It appears that, parallel to our familiar flat world, a multi-dimensional non-linear world influences our world.

The Information Age has opened a window to the other worlds outside our "standard" physical world, enabling us to reach dimensions that we could not access before, so we need to learn to live with this and develop suitable vehicles to "maneuver" between these new horizons and our traditional world.

From the time of Isaac Newton, we have harnessed reality to mathematical, linear principles that enabled us to develop technology and live in a world that was comfortable for us. In the military sphere at various periods, armies formulated principles based on ease of control and adapted to the resources at their disposal, from Sun Tzu's reliance on cognitive leadership to European armies based on geometric formations.

Today's world is making the transition to reliance on digital information and the technological developments stemming from it. What was sufficient in the Napoleonic Era and the wars that followed is no longer enough. Reality is becoming more complex; as better tools are developed, we may observe and analyze many more aspects that influence our functioning and achievements. At the same time, we are becoming aware of a diminishing ability to describe and influence with current tools. The world opening before us today is far more complex and if we acknowledge its existence and our limits, we will need to create new vehicles to cope with it and operate in the new reality.

Even if we do not yet understand it fully, the digital world provides us with tools to open the door and step in. The digital world is multi-dimensional, unlimited, and infinite, and accords time and space an entirely new meaning. Our traditional perceptions of weight, distance, and volume are replaced by concepts involving time, resolution, and availability. This world does not eliminate the physical one in which we live but enriches it immeasurably, creating new perspectives.

Organizing Command and Control on the Non-linear Battlefield

This chapter does not deal with the operational side of warfare, so we will focus on information as the heart of any command and control system. The physical structure of an information center is irrelevant; it might be a single computer terminal or comprise a large number of computers and personnel. Its purpose is to gather information, process it, and apply it, making it available for its own purposes and those of additional users. Information centers exist at every level connected by networking systems.

Since command on the tactical level operations is hierarchical, a different type of interface is necessary between information systems and physical components. In the Information Age, the volume of information at the commander's disposal is much greater than he can handle, so technical solutions are necessary that receive, process, and distribute greater amounts of information. However, not only volume has changed; access to supportive centers of power and hubs of information have undergone a "population explosion." In addition to traditional "pure" information sources, net-centric warfare connects other means of support on the net as well; thus, the amount of information is enormous. A professional, mission-oriented cellular structure is required, a crucial means of allowing the commander to fully enable his unit to operate efficiently. We had in the past, and still possess, radio networks dedicated to connecting units and services, but they are restricted in range and volume, whereas computerized networking can connect an unlimited number of information providers, and cellular organizations can engineer who can do what, when and about what.

Based on such cellular networking, a staff can offer the commander much more relevant and timely information than the present system is able to do. Moreover, it creates the ability to separate the control from the command function and let the staff become an operational arm in the commander's service, thus releasing them to carry out their main function: command.

Mission control can come to the aid of this complex interface. As an organizational framework, this approach comprises autonomous centers that make decisions and manage operations in their fields of expertise and authority for achieving their defined objectives. These centers are supported by others—each with its own field of expertise, whether firepower, intelligence, logistics, etc.—that are integrated into the mission's cellular networking system to achieve objectives.

The element uniting all these centers is an information system needed to accomplish the mission. Such a battle management system interfaces hierarchical command; the tactical unit, with supportive centers of power that are not necessarily under the sway of a hierarchical system, but with information and capabilities that are temporarily assigned to it. This exists today to a certain degree, but when it is possible to integrate powers based on up-to-date information, a fighting formation will be reinforced by truly multi-dimensional support.

These capabilities have existed in the past, but changes in relevant time dimensions, informational requirements, and scope have led to improved command, control, and management systems face-to-face, but doing so demands thinking outside the box.

An organizational distinction between command, control (responsibility and authority), and staff work (preparation, supervision, management) reduces pressure on the commander and frees them to concentrate on the mission. Simultaneously, it maximizes the range of services that the staff and headquarters can and must provide to this end. Mission command imposes a greater burden on the higher echelons and supportive headquarters, while lightening the load on the commander's shoulders and providing them with what is necessary, including the freedom to make decisions.

According to this approach, the accepted separation of strategic, operational, and tactical loses much of its relevancy. If the system develops a reliable, supportive net-centric system that is accessible to all—based on information and an authority control plan—it will be possible to manage a campaign on two levels only: the tactical-contact level and the strategic-campaign level that comprises the commander's intentions and apportioning of resources.

Changing Priorities in Staff Work

Staff work relies on two technical frameworks:

1. Information management systems dealing with collection, processing, and distribution; integrating mechanized systems; information management teams; and content experts;

2. Time and means management systems dealing with planning and coordinating processes of resource allocation based on the current plan; integrating and preparing the next battle plan for the commander.

Successful professional management of these two supportive systems is a prerequisite for effective continuous fighting management. Staff work must be flexible and staff headquarters organized according to mission command principles. Mission-oriented staff teams can centralize and decentralize according to need, and networking information systems can enable the transmission of necessary data and its products at any time or place. The headquarters is one of the nodes on the mission's net-centric system.

Technological capabilities enable information to be retrieved from remote dedicated centers that are available as information sub-networks to all users.

The Approach to Command in Net-centric Warfare

The approach presented here rests on the assumption that, in net-centric warfare, the architecture of the network system must support mission command from the outset as the central model for command and control and battle management. According to this approach, command and control are not limited to the ingenuity and expertise of a local commander. Rather, the entire system is adapted to support mission command-style battle management. "Information pulling," rather than higher echelons supplying information as in a hierarchal system, enables "clients" to follow the mission control concept "support command and control." This means that the entire system will be changed based on a variety of resources and information teams.

The architecture of a system that can support mission command is based on an operational control format (JP6) that is fundamentally rigid and rarely changes, around which a dynamic infrastructure is built of computerized information management functioning in a tactical control (JP1) format.[1] Nevertheless, implementation lags behind since we are slow at making a cognitive leap forward.

To this end, it is necessary to determine the components that can transform today's standardized hierarchical command and control systems into mission command-centered systems. This involves two operational dimensions: a hierarchal chain of command and net-centric support.

1   Tactical control provides sufficient authority for controlling and directing the application of force or tactical use of combat support within the assigned mission or task (JP1).

In the information era, managing and commanding fighting takes on a new form. Potential and optional courses of action are created for "the next step," thus easing the efficiency of activating forces and resources for continuous fighting. Structures of this type enable maximum utilization and economy of forces and means, lessening overload on headquarters, exploiting time and space and improving versatility.

When the command and control system enables each commander having direct access to information and when they have direct contact with centers of power regarding combat support and combat support services, the time span necessary for activation and coordination will be considerably shortened. The commander's ability to take the initiative increases, and they can make decisions with minimal dependence on hierarchies.

The hub of the system is the hierarchical command and control system, as a unified command and objective are sacred values of battle management. It should be kept thin and effective, the net-centric part should be wide and flat, and control over the flow of information should done by access control.

The new technological advances enable the commander to utilize management capabilities that are not necessarily in his geographical area or under their command; therefore, two parameters are necessary in order to integrate systems: effective range and precision. This holds true for both information and combat materials. The framework in which that data center functions is irrelevant; commanders can reach them at any time and get the information they are looking for from a non-hierarchical network. The space dimension expands.

Net-centric warfare also brings about significant changes in the time dimension. Data concerning intelligence or activating supportive resources is constantly available, meaning that combat can be continuous, so time becomes a weapon system. According to the OODA loop theory developed by John Boyd (2007) based on his experiences as a pilot, preempting the enemy by means of efficient exploitation of time becomes a central component in this new warfare.

When analyzing a situation, commanders tend to analyze quantities based on a hierarchical approach. Conversely, network-based activity and mission command grounded in enhancing power ratios involves the force's actual performance at the point of contact with the enemy at a particular time. These power ratios are constantly in flux, without necessarily being determined by statistics regarding forces and resources appearing on staff charts. In order to use this as a force multiplier, headquarters and staff must be adapted to function in net-centric warfare formats and officers must trained for multi-tasking that is coordinated by a unified command.

The implications of the information era and mission command for battle doctrine include data availability, increased availability of data-dependent resources, such as weapon systems and logistics, and breaking down bureaucratic and hierarchical barriers to networking communication. All of these enhance the planning of command and management capabilities and enable a cognitive breakthrough into a non-linear world that was previously concealed from view.

The Commander and the Staff

In the information era, the commander is required to increase his capabilities, as compared to those required by a hierarchical-linear format. With more resources given to them, they are expected to plan and execute their mission their way. By employing net-centric warfare, more staff support is given to commanders by way of information availability. The commander will have a wide array of supportive data and resources at their disposal. Resource usage depends on them. Instead of waiting for data to be sent to them, it will now be available to the commander from the beginning, and they can retrieve it at their convenience. Staff work must adapt to a changing hierarchy and a sped-up tempo, and commander and staff must adapt to these new capabilities. Consequently, headquarters will undergo fundamental changes; it will function no longer solely as the commander's assistant but as a fighting unit in its own right, activating resources that are at times not under direct control.

The staff will continue to aid and support the commander with far more resources, not necessarily through geographical proximity or direct contact. Doing so will speed up action and require different (better) qualifications for staff officers, different decision-making procedures, and highly qualified commanders. Too many variables in the system place us inside the non-linear spectrum; thus, we should take one step at a time. Assignments and mission work should be broken down into cells of a size that the staff officer can handle in terms of pressure; the network should be arranged to function according to mission-based net-centric cells.

Command and Control as a Weapon System

In the information era, command and control systems change from supportive tools to weapon systems. As the fighting tempo speeds up and the potential number of parties involved in the fighting increases, battle management becomes a central component in deciding outcomes.

Face-to-face fighting results directly from a plan that the execution of which must create local superiority in time and space. The organization of linear warfare relies on a plan with clear end-state objectives, by either main and secondary efforts or their variations with a preliminary set of forces. In non-linear warfare, the road to end-states is convoluted, requiring a number of small steps at each phase. The transition from one phase to another cannot be predicted in advance; rather, it results from the previous phase based on success and situation development. Forces are deployed to perform dynamic, flexible fighting. Thus, the command and control system will have to cope with dynamic events on the battlefield, set the pace, develop local and temporary supremacies, and ensure that the way to the final objective is as short as possible.

The command and control system must support rapid, continuous fighting according to the abilities and needs of the operational forces. It is reinforced by the capabilities and support of "every capable means" available within effective range or that holds effective data. When information is available and accessible and when there are technological solutions for every relevant range and target, two problems remain: the first is an organizational structure and deployment enabling the dynamic and efficient activation of forces, and the second is a command and control system with capabilities and skills for initiating multi-dimensional operations. Mission command makes this possible.

Applying these systems to the new battlefield opens up new possibilities for practicing the art of war, taking us away from numerical calculations based on human limitations toward new insights. Mission command is based on these insights and will bring the art of war into the Information Age of the twenty-first century.

Doing so demands quality commanders who, in addition to their basic skills and expertise in deploying resources, will be more than just military leaders. They will need to be "artists of war" capable of initiating, inventing, taking risks, and controlling events.

Mission Control, the Commander, and the Headquarters

Data processing occurs in three major phases: facts that become information, which in turn becomes knowledge. This is a never-ending process, since every piece of information contains more than basic facts, while knowledge always reveals something new. It is non-linear process from the outset, as information flows in constant motion and has value insofar as a person or a program derives new insights from it.

Information management systems have always been present in the military to support optimal decision-making, constituting the major force promoting staff work and providing the basis for its functions. In the information era and according to the precepts of mission command, they have an additional goal: to generate information that is at the disposal of commanders at different levels for decision-making purposes. In the past (and still today), such databases were for the exclusive use of the local commander, but in the new era, they have broken through barriers; databases are available to anybody, subject to security clearance. The system is no longer available to subordinates according to the judgment of their superiors but instead shared by commanders and junior officers at all levels according to professional needs.

Prior to the digital era, staff control centers were constructed around data, but they flowed in one professional, authoritative direction. Communication channels connected similar bodies: operations to operations, firepower to firepower; logistics to logistics, and maintenance and maintenance. These complexes collected relevant data in cells (operational, medical, artillery, engineering, logistics, etc.). The information was then processed and distributed to professional bodies for processing and completion and was then transferred to a combined management center which then disseminated it to those commanding the battle. This system had the one disadvantage of not considering "foreign" influences; they were only included in the equation later on in the process.

Conversely, digital systems function according to topical categories rather than professional or authoritative ones. Any interested party might access accumulated data derived from various sources. Such a system enables control of data flow, its processing and transmission to any part of the system. Promoting data according to topic makes it possible to shorten processing time and manpower and, especially, to improve data processing and distribution among consumers. Of course, every professional or command network maintains a hierarchy ensuring reliability and quality, but information is readily available to all network participants.

Partnership with the Digital World

Digital systems for managing and processing information are based on computers at various technical levels, but computers cannot think for themselves. Computerized command and control systems and management systems are aimed at assisting commanders with planning, command, supervision, and control of fighting. In the three-part process described above (data-information-knowledge), the mechanized component plays a major part in the first stage of collecting and storing data; it plays a smaller part in the second stage that involves processing facts into information, although it will still support sound planning procedures. However, with the transition from information to knowledge, human insights will dominate, with digital technologies playing a supportive role. In other words, the computer is managed by means of previously determined protocols and operated by expert analysts who process data according to the needs of commanders. Winning the battle will always ultimately rely on the commander's cognitive skills.

Information management serves staff work and planning, as well as the command and control system; each possesses its own requirements, but they act in harmony. The structure of staff headquarters and decision-making processes must ensure that this harmony is maintained at all times.

Doing so relies on decentralizing functions at headquarters in the form of data centers from which the necessary information is derived. The commander deals with planning and commanding the battle, and the staff partially supports them and partially administers a data center with a wide range of information. This structure already exists today, but scope and work methods differ. Networked communication makes it possible to concentrate and process information in professional centers and leave it to the staff to assemble, process and distribute what is necessary for the mission. Thus, the staff takes on a new task, namely, data management.

Data Management Systems

Every data management system focuses on two spheres: firstly, collection and processing that is done in specialized centers, and, secondly, support for decision-making done at command and control centers. Between the two are information networking channels. If we succeed in creating computerized systems that intensify processing capabilities to the level of artificial intelligence and the organizational structure is upgraded to accommodate these capabilities, we will be able to concentrate human effort on activities that machines cannot perform, consequently reaching new levels of quality and better exploiting the resources at our disposal.

As a surfeit of information can create severe difficulties in utilizing information and ensuring that it is available to those seeking it, data management systems must be efficient and reliable. Such problems may be solved technologically by minimizing the scope of a single professional network center and managing mission- or issue-dedicated networks, which operate according to mission command as well. This means networks are assembled and disassembled according to need so that the higher level can support lower-level operations.

Two types of networks are necessary for this arrangement: information networks for supporting immediate needs like command, control, intelligence, and fire; and supporting networks for operating resources for force logistics and deployment, monitoring, and medium- and long-term pre-planning. In order to promote staff work, these two networks must be in full communication with each other. Mission command enables the commander to tailor specific connections between the two systems; although the second supporting system is not under their command, they can authorize access to it according to their operational planning.

Data Management as the Basis for Mission Command in Control Centers

All information takes on a different meaning according to consumers' purposes. For instance, information regarding enemy deployment allows intelligence to estimate enemy activity, artillery to plot targets, and operations to plan maneuvers and fire power. How effective information is depends on end users and the degree to which they receive it in a form that serves their needs within the period necessary for them to make decisions.

Below, we describe these two elements: information centers and information available to the decision makers who require it.

- Information centers—the system that receives, stores, processes, and distributes information.
  - ▸ Large amounts of information constantly arrive from a variety of sources.
  - ▸ Information is categorized according to relevance, purpose, and urgency for potential end users.
  - ▸ There may be users who require all or part of the same information.
  - ▸ Processed information takes on new meaning in a never-ending process.
  - ▸ The usefulness of information changes according to the use made of it.
- Command and control centers
  - ▸ Many users require information in order to reach decisions.
  - ▸ Users require the information in a format that supports their decisions.
  - ▸ Processing and presenting information is multi-directional. Each individual piece of processed information is likely to create the need for additional information.
  - ▸ Processing and presenting information is non-linear. It is

subject to constantly changing influences, implications, and intensities.

Since different users need information that is processed in different ways, the manner in which information is presented is a critical factor in determining its value. Thus, the system must be flexible, simultaneously available to a number of consumers and user-friendly.

Staff work must focus on preparing information, presenting it, and managing operations with the support of digital systems that improve their output and shorten the time required to produce it. These capabilities do not alter anything concerning the commander's insights and decision-making; their responsibilities remain very hierarchical.

Points of Reference

Information is the driving force behind operational management.

- The quality of information and its availability play a major part in successfully achieving an operation's objectives.
- The digital system contains huge amounts of information from countless sources, thereby diverging from the boundaries of the former professional, hierarchical context.
- Digital information systems afford supervision, management, and dissemination of information that cross hierarchical boundaries and afford operational forces and decision-makers direct access.
- Military operations have a single commander, while unity of objective and command are a cornerstone of warfare; the decentralization of information must not threaten this principle.

Basic Assumptions

- There is more than one reality and both creating and managing the one that is relevant to a particular mission and setting depends on the quality of information and its management according to planning and deploying forces suited to that mission.

- A commander activating their forces by means of sub-commanders according to a hierarchical command system manages military operations.
- The command center having the ability to manage information allows it to exercise command and control over all the forces under its command, thus constituting a mission center of gravity in commanding operations.
- While the commander manages fighting on a hierarchical axis, headquarters functions as a multi-dimensional network equipped to manage information, supervise activities, and control their execution.

Repercussions

- Although the role of headquarters in supporting the commander has not changed, it has acquired new capabilities. New needs have developed on the battlefield, enabling and demanding headquarters to broaden its interests and activities beyond previous limits.
- The influence of a headquarters' fulfilling all its responsibilities transforms it into a weapon system and force multiplier.
- A combined and coordinated effort of a hierarchical command (the commander and their assistance staff) and a network command and control system (headquarters) is likely to result in more efficient capabilities exploitation of the forces and enhance effective battle management.
- Making the distinction between the two tasks of command and control demands organizational adjustments and work processes that will bring about full realization of capabilities, while every organization is suited to its specific tasks and role.
- The commander and the command team comprise operational management.

- The control center comprises planning, supervision, control, and activating efforts and resources as part of the commander's operational plan.

Technology and Mission Control

In theory, weapon systems are developed according to operational needs, but, in fact, military industries develop according to the technologies at their disposal then market them to the army, which follows their lead. Arms development stems from a problem requiring a solution. Warfare in a mission command format demands the characterization and development of war resources suited to its fighting principles which industry cannot develop.

According to battle doctrine, the professional field of military technological is relatively narrow regarding each individual weapon, and as expertise increases, this trend is even more evident. However, the battlefield comprises forces woven together into a multi-dimensional whole. Thus, the mutual influence of each resource on others is just as important, or more so, than its individual capabilities.

As the army starts thinking in terms of a multi-dimensional battle environment and net-centric warfare, it must also consider multi-dimensional technological solutions and net-centric technologies. The information era allows this to happen. The new era makes it possible to activate weapon systems toward a common goal by any organization and from any geographical location, especially when they are developed on an interface combining their capabilities in one management and command framework.

It thus becomes obvious that, alongside the specific capabilities of each system, operational requirements demand placing emphasis on interfaces containing supplementary support systems that are not dependent on manufacturers' original, unique technologies.

The Influential Factors Shift

The two major factors influencing warfare are time and space and how they are utilized; this constitutes the core of all military activity. Neither time nor space is passive or dependent on force majeure; they are at the disposal of the commander who is responsible for creating temporal and spatial superiority by wisely utilizing environmental conditions and the resources and forces at their disposal. A large number of modern technological developments are directed at improving temporal and spatial management.

Time

The ability to maximize our strengths and pre-empt the enemy by exploiting our capabilities is a key factor in achieving decisive superiority. This is true at the lowest tactical level; "looker-shooter" is a good example when one is dealing with disappearing targets, as well as on the operational level. In the Six-Day War, the I.D.F. achieved a decisive advantage on the ground by quickly and effectively deploying forces, bringing victory on the Egyptian front regardless of the size and quality of the forces in the field. Conversely, the hesitation and slow pace displayed by the I.D.F. in the Second Lebanon War and in the Gaza Strip in subsequent years resulted in a stalemate, despite the relative size of the forces, even when they were clearly in Israel's favor. These examples demonstrate that time is a non-lethal weapon that, when wisely managed and exploited, can be a decisive factor in battle. The time element is under the commander's control. With the assistance of information systems, the temporal gap between receiving data and utilizing it is constantly narrowing.

Space

Warfare is waged against an enemy and deployed in four dimensions: land, air, sea, and information. The four are not static; in the information era, the battlefield is likely to be urban and a significant proportion of the battle area will be located deep in enemy territory.[2] In most cases, there will be no clear front line; fighting will be multi-dimensional with a 360-degree range.

The fighting area is plotted out at the planning stage, when it is determined where and how we choose to maneuver and fire and from where we can decisively influence the development of the operation, whether by occupying territory or by gaining control over enemy activity. The selected battle arena is not necessarily static and fixed; instead, it might be flexible, mobile, and dispersed over a large area.[3] The result of careful strategic planning is likely to determine the outcome of the battle before it begins.

According to non-linear battle principles, the battlefield's importance changes according to the situation on the ground, while planning involves a large number of small steps and dynamic battle management.

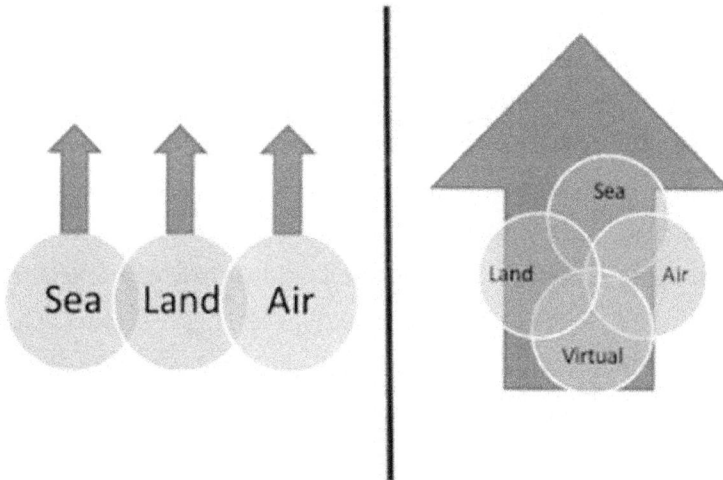

Figure 10: Unified battle

2   "Depth" means the distance from our lines and it is not fixed, but determined at all times by our ability to act in it. The greater the range of weapons, intelligence and command and control, the broader will be the proportion of face-to-face fighting and deep zones.
3   There might be a number of systemically controlled, but geographically and spatially scattered, fighting areas.

Information Warfare and Multi-Dimensional Battles

The information—or virtual—sphere has developed slowly from the invention of the radio and other media use for control to digital information systems. They have become a battle space in themselves when every device gives rise to a counter-device, with cyber technology being a striking example of this. The information environment has become equally dominant on land, in the air, and at sea. It exists as a stand-alone battle environment (including irregular warfare) but, at the same time, is active parallel to the others, completing and enhancing them by adding another dimension altogether.

Information is integrated into every aspect of warfare. It has many branches, with a prominent one being cyber warfare, but this is neither unique nor always the most significant. Electronic and electromagnetic warfare participate in the information war, either on their own, combined with other elements, or as backup for land, sea, and air battles.

Information is a vital element of all weapon systems used by the army, in the form of information-based automatic or autonomous systems: data automatically programmed in advance or data that is processed in real time by autonomous systems. The involvement of humans at various levels in these systems constitutes the cooperation between man and machine and fighting systems.

The significance of the information era is restricted not only to the appearance of a new weapon system or fighting arena but also to changes in attitude toward organizing and managing battle arenas and warfare in general. Networking gives us net-centric warfare that opens up new opportunities for organizing warfare based on operational characteristics and availability of information, as well as on interfacing as a vital element in organizing forces, to the point where digital networks may be considered non-lethal weapon systems.

Organizational hierarchies that are constructed due to difficulties in sharing and controlling information are no longer necessary. It is possible to include and combine forces and resources based on shared information and activate them when spatially distant from one another, while still working in harmony under one commander and toward one objective.

The Integrated Battle as a Condition for Mission Control

The concepts of integrated battles, cooperation, and combined efforts must all be examined afresh in the information era. These are multi-dimensional integrated battles which need integrated combat forces, regardless of their origin or heritage. When weapon systems are based on interfaced technologies, the battlefield becomes unified. Battlefield organization must be mission oriented and managed according to mission command due to its being complex rather than resource-centered.

As far as possible, technological resources must be developed toward cooperation. Although all technical devices provide limited solutions, their combined performance on the battlefield is a higher priority than is their individual capabilities.

The starting point of battle planning is the commander's intentions and the allotment of resources for the mission. From that point on, managing and controlling the battle is in the hands of sub-commanders with assigned missions and ad hoc battle teams responsible for deployment, exploiting advantages and completing missions.

Deployment ranges are no longer limited but now lie in the ability to combine and share information. Task forces are efficient as long as their activities are performed within a relevant period[4] and their output impacts the mission at the time and place required. The commander can make a major contribution to activating his forces by the wise exploitation of time and space. Taking the initiative and controlling the forces' concentration and dispersion enable him to confound the enemy's capabilities and gain supremacy.

In the information era, the concrete battle area expands far beyond

4   Relevant time is that in which a necessary activity carried out within the framework of a mission.

face-to-face fighting. Systems influencing tactical fighting (and higher) are likely to be found outside the range of the weapon systems in the hands of the fighting force, but they might have an impact through direct or indirect involvement in a short time span or even immediately. The ground battle not only overlaps with low-altitude air space but also with an invisible information environment.

According to the mission command format, battle organization is ideally based on decentralized reserves of forces and resources, maximum utilization of all capabilities, ad hoc battle teams that can easily change direction and effort, flexibility of deployment, and decisive superiority in contact, all of which may be graphically described as "fighting from the depth to the depth."

Repercussions for Battle Doctrine

Battle doctrine combines functional and operational principles. The human element comes to expression through insight and leadership that maximize capabilities in developing circumstances and dynamic environmental conditions. The level of excellence attained depends on the quality of battle organization and the availability of resources for commanders to activate.

Mission Control

In the information era, it enables the commander to receive and pass on much more data than was possible at the beginning of the present century. Efficient data management and mission command and professional network activity open up a wider range of choices before the commander than ever before. More information aids the commander in making decisions with much less dependence on the higher echelons and affords him tools to function in a true mission command framework.

The shorter time span necessary for making decisions and carrying them out also renders adherence to mission command principles the preferred command style. A flexible approach to fighting and integrated support systems will improve the commander's ability to realize these principles to the fullest.

Flexibility

This means the ability to combine and organize teams; decentralize and centralize forces and resources unrelated to physical location or origin, enabling the commander to change their mind; and deploy and concentrate forces and resources without jeopardizing fighting power.

Their wide range of choices and ability to carry them out generate greater functional flexibility. Mutual interfaces, a common language, unity of mission, and command will provide any fighting organization with the necessary flexibility to perform its missions.

Initiative

When the commander is supported and strengthened by means that are not under their direct control, they can operate outside the box, knowing that the system will provide the necessary additional resources for the initiatives. When they sees the possibility of achieving local, or even temporary, superiority, an energetic commander can exercise initiative, exploit circumstances, solve problems, and lead the battle according to their own individual style.

Deception

The Information Age creates endless opportunities for deception, trickery, pretext, and other surreptitious means of gaining supremacy.

Influences on the Organizational Framework

The information era has broad implications for the organizational framework. It leads to a shift in the power and importance of various components.

Numerous and varied solutions exist today that are far more unique than in the past due to their being based on expert technological systems. These systems are limited to the applications for which they were developed. While their performance is vastly improved, their ability to be combined is significantly limited. Pre-programed machines now perform what human beings formerly instinctively carried out, and people must adapt themselves to this situation. It is impossible to convince a computer to make changes if it is not programmed in advance to function with change. Thus, simplicity is necessary when activating a wide range of resources. Battle management must be based on four fundamental principles: a common language; simplicity; a clear, understandable functional approach; and ongoing organizational flexibility. This management will enable the following:

- Freedom of action in conditions of troop saturation and complex areas
- Flexible deployment at any width and depth that is necessary, while reserving the ability to concentrate decisive effort at the selected time and place
- Economical use of forces through modular structures and shared interfaces
- Flexible fighting formation structured according to immediate ongoing and changing needs on the battlefield
- Continuous action with the necessary decisive power
- Organizing units/formations for battle

The information era enhances the ability to organize teams and resources at lower levels. More flexible combat organizations are required that maintain the ability to concentrate large, powerful forces when necessary for decisive action.

There remains the need to concentrate effort on strategic objectives and to concentrate forces and means. However, an enlarged battle area and decentralized fighting, together with the concept of battles "from depth to depth," demand a different mindset to deal with the problem of how to exploit strengths and excellence to gain supremacy and decisive victory.

Large maneuvering bodies are necessary for in-depth resolution of fighting, but "large" refers to the power exerted on the enemy, not on concentrated organized numbers. The information era has brought about the integration of autonomous and automatic systems into warfare, enabling us to deploy the best possible synergetic combinations of resources and a wide range of activities that do not detract but rather improve the quality and flexibility of performance.

While maintaining maximum precision of warheads, the greater effectiveness of intelligence and firing capabilities lessens the need for large concentrations of forces and resources in advance as the solution to every difficulty on the battlefield. Available weapon systems may be widely decentralized but still readily available to the commander for activation.

Fighting is a combination of face-to-face fighting and supportive resources for a particular mission. The combined mission command battle takes place simultaneously at every range and location in which it is possible to combine data and weapon systems in a united mission under one commander.

Continuous fighting constitutes one way of exerting pressure on the enemy, disrupting the rationale behind its deployment and thwarting its ability to respond. Continuous fighting is achieved by mobility and transportation of means and forces. Deploying a range of resources and forces from depth to depth enables continuous fighting without the need to concentrate large forces on the battle line. The battle for supremacy in continuous fighting is one of the crucial elements of any military attack. Employing the information virtual battle space as an integral part of the campaign supports the battle continuation and the enemy systems disruption.

Organization based on these concepts is valid for every fighting level according to conditions and actual needs.

The Significance of the Above

- Combined battles and mission control require a simple, basic organization; a clear, structured battle order; a common language; and technological interfacing among resources and systems.
- Integrating warfare into the information sphere and from it to other fighting arenas is the key to maximum exploitation of information technology and synergy among all the forces and resources active in a mission.
- A mission command networked fighting organization will replace former hierarchical corps-centered frameworks based on limited resources.
- Improved tools for data management, new staff procedures, and control systems based on information era capabilities are all needed in order to support a net-centric warfare organizational framework.
- Commander training must accommodate changing conditions; new ways of control must be developed, along with mobility and transportation of forces and resources for multi-dimensional depth-to-depth fighting.
- Databases are needed that provide optimal solutions for every level, a kind of super-network that encompasses professional and command and control networks. These systems must be flexible and subject to change ad hoc, adding and removing users, and sorting and distributing information according to the needs of the mission.
- Data management has become a crucial element in modern warfare.

Brig. General (Ret.) Gideon Avidor recruited to the I.D.F. in 1957 to the Armored Corps. He commanded Tank Company, Battalion, and Brigade. He participated in the Six-Day War as Tank Brigade G3; in the Yom Kippur war, he was the 252nd Division G3. During the Second Lebanon War, he was the Chief Armored Officer. He also commanded the School of Armor, was the DY Armor Corps Commander, the Israeli Defense Attaché, and the Ministry of Defense representative to Singapore, Australia, Papua-New Guinea, New Zealand, and the Philippines.

He graduated from the U.S. Army Command and General Staff College and the I.D.F. National Defense College. He holds a B.A. in History and an M.A. in Urban Geography.

Appendix A

Terminology

close support—The action of the supporting force against targets or objectives that are sufficiently near the supported force as to require detailed integration or coordination of the supporting action.

combat information—Unevaluated data, gathered by or provided directly to the tactical commander, which, due to its highly perishable nature or the criticality of the situation, cannot be processed into tactical intelligence in time to satisfy the user's tactical intelligence requirements.

combat power—The total means of destructive and/or disruptive force which a military unit/formation can apply against the opponent at a given time.

command and control system—The facilities, equipment, communications, procedures, and personnel essential to a commander for planning, directing, and controlling operations of assigned and attached forces pursuant to the missions assigned.

command and control (C2)—The exercise of authorityy and direction by a properly designated commander over assigned and attached forces in the accomplishment of the mission. Command and control functions are performed through an arrangement of command and control systems.

commander's intent—A concise expression of the purpose of the operation and the desired end state. It may include the commander's assessment of the adversary commander's intent and an assessment of where and how much risk is acceptable during the operation.

mission command—The preferable command method is mission command. It assumes that every commander is best suitable to perform the mission at their level, in their sector, and with their forces. The higher-level commander dictates the mission goal and what it envelopes (resources and restrictions). The subordinate commander needs to decide and act to achieve their part of the mission. In certain conditions, mission command is replaced by detailed command in which the higher echelon dictates the methods on carrying the mission as well. (Ground Forces Command, *Tactical leadership at the Ground Forces*, 2012, p.22)

mission statement—A short sentence or paragraph that describes the organizations essential task (or tasks) and purpose. A clear statement of the action to be taken and the reason for doing so. The mission statement contains the elements of who, what, when, where, and why, but seldom specifies how.

mission type order—(1) An order issued to a lower unit that includes the accomplishment of the total mission assign to the higher headquarters. (2) An order to a unit to perform a mission without specifying how it is to be accomplished.

mutual support—The support which units render each other against an enemy, because of their assigned tasks, their position relative to each other and to the enemy, and their inherent capabilities.

graphic order—Operation order in which major parts appears in graphic format rather than in words. (I.D.F. *Lexicon*, 1998, p. 510)

standard operating procedure (SOP)—A set of instructions applicable to those features of operations that lend themselves to a definite or standardized procedure without loss of effectiveness.

Abbreviations

AO	area of operations
AOI	area of interest
AOR	area of responsibility
ATTP	Army tactics, techniques, and procedures
C2	command and control
CCIR	commander's critical information requirement
CIE	collaborative information environment
CO	cyberspace operations
COA	course of action
COG	center of gravity
CONOPS	concept of operations
CONPLAN	operation plan in concept format
COP	common operational picture
IO	information operations
ISR	intelligence, surveillance, and reconnaissance
NETOPS	network operations
OPCON	operational control
OPLAN	operation plan
OPORD	operation order
TACON	tactical control
TF	task force
TST	time-sensitive target

www.ingramcontent.com/pod-product-compliance
Lightning Source LLC
Chambersburg PA
CBHW021548210326
41599CB00010B/352